Contents

Content Guidance

Questions & Answers

Getting the most from this book

Examiner tips

Advice from the examiner on key points in the text to help you learn and recall unit content, avoid pitfalls, and polish your exam technique in order to boost your grade.

Knowledge check

Rapid-fire questions throughout the Content Guidance section to check your understanding.

Knowledge check answers

1 Turn to the back of the book for the Knowledge check answers.

Summary

Summaries

- Each core topic is rounded off by a bullet-list summary for quick-check reference of what you need to know.

Questions & Answers

Exam-style questions

Examiner comments on the questions
Tips on what you need to do to gain full marks, indicated by the icon ⓔ.

Sample student answers
Practise the questions, then look at the student answers that follow each set of questions.

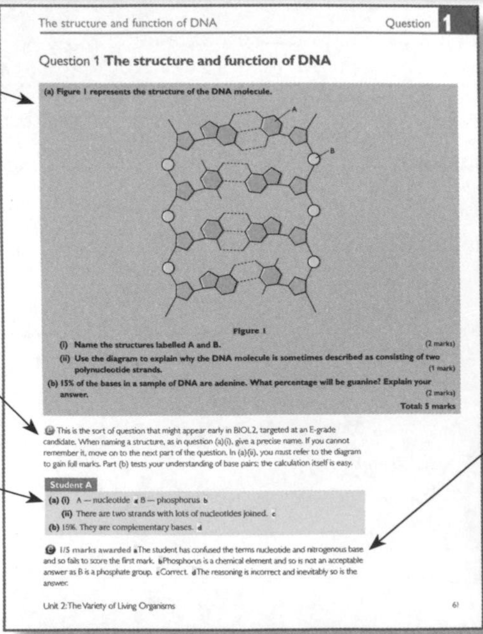

The structure and function of DNA — Question 1

Question 1 **The structure and function of DNA**

(a) Figure 1 represents the structure of the DNA molecule.

Figure 1

(i) Name the structures labelled A and B. (2 marks)

(ii) Use the diagram to explain why the DNA molecule is sometimes described as consisting of two polynucleotide strands. (1 mark)

(b) 15% of the bases in a sample of DNA are adenine. What percentage will be guanine? Explain your answer. (2 marks)

Total: 5 marks

ⓔ This is the sort of question that might appear early in BIOL2, targeted at an E-grade candidate. When naming a structure, as in question (a)(i), give a precise name. If you cannot remember it, move on to the next part of the question. In (a)(ii), you *must* refer to the diagram to gain full marks. Part (b) tests your understanding of base pairs; the calculation itself is easy.

Student A

(a) (i) A — nucleotide **a** B — phosphorus **b**

(ii) There are two strands with lots of nucleotides joined. **c**

(b) 15%. They are complementary bases. **d**

ⓔ 1/5 marks awarded **a** The student has confused the terms nucleotide and nitrogenous base and so fails to score the first mark. **b** Phosphorus is a chemical element and so is not an acceptable answer as B is a phosphate group. **c** Correct. **d** The reasoning is incorrect and inevitably so is the answer.

Unit 2: The Variety of Living Organisms 61

Examiner commentary on sample student answers
Find out how many marks each answer would be awarded in the exam and then read the examiner comments (preceded by the icon ⓔ) following each student answer. Annotations that link back to points made in the student answers show exactly how and where marks are gained or lost.

About this book

This guide will help you to prepare for **BIOL2**, the examination for **Unit 2: The Variety of Living Organisms** of the AQA A-level Biology specification. Your understanding of some of the principles in Unit 1 may be re-examined here as well.

The **Content Guidance** section covers all the facts you need to know and concepts you need to understand for BIOL 2. In each topic, the concepts are presented first. It is a good idea to get your mind around these key ideas before you try to learn all the associated facts. The Content Guidance also includes exam tips and knowledge checks to help you prepare for BIOL2.

The **Question and Answer** section shows you the sorts of questions you can expect in the unit test. It would be impossible to give examples of every kind of question in one book, but these should give you a flavour of what to expect. Each question has been attempted by two students, Student A and Student B. Their answers, along with the examiner's comments, should help you to see what you need to do to score a good mark — and how you can easily *not* score a mark even though you probably understand the biology.

What can I assume about this book?

You can assume that:
- the basic facts you need to know and understand are stated explicitly
- the major concepts you need to understand are explained clearly
- the questions at the end of the guide are similar in style to those that will appear in the BIOL2 unit test
- some of the questions test aspects of *How Science Works*
- the answers supplied are the answers of AS students
- the standard of the marking is broadly equivalent to that which will be applied to your answers

So how should I use this book?

The guide lends itself to a number of uses throughout your course — it is not *just* a revision aid. You could:
- use it to check that your notes cover the material required by the specification
- use it to identify your strengths and weaknesses
- use it as a reference for homework and internal tests
- use it during your revision to prepare 'bite-sized' chunks of related material, rather than being faced with a file full of notes

You could use the Question and Answer section to:
- identify the terms used by examiners in questions and what they expect of you
- familiarise yourself with the style of questions you can expect
- identify the ways in which marks are gained as well as how they are not gained

Develop *your* examination strategy

Just as reading a book about pianos will not make you a good keyboard player, this guide cannot help to make you a good examination candidate unless you develop and maintain all the skills that examiners will test in BIOL2. You also need to be aware of the type of questions examiners ask and how to use the mark allocation to answer them. You can then develop your own personal examination strategy. But, be warned, this is a highly personal and long-term process; you cannot do it a few days before the unit test.

Things you *must* do

- Clearly, you must know some biology. If you don't, you cannot expect to get a good grade. This guide provides a succinct summary of the biology you must recall and understand.
- Be aware of the skills that examiners *must* test in BIOL2. These are called assessment objectives and are described in the AQA Biology specification.
- Understand the weighting of the assessment objectives that will be used in BIOL2. Examiners have designed BIOL2 with the approximate balance of marks shown in the table.

Assessment objective	Brief summary	Marks in BIOL2
AO1	Knowledge and understanding	33
AO2	Application of knowledge and understanding	32
AO3	*How Science Works*	20

- Use past questions and other exercises to develop all the skills that examiners must test. Once you have developed them all, keep practising to maintain them.
- Understand where in BIOL2 different types of questions occur. For example, the last two questions will always be worth 15 marks each. One will test your data handling skills (AO2) and end with a 6-mark part requiring continuous prose and testing mainly AO1. The other (the final question) can be based on content selected from anywhere within the AS specification. This will test your ability to critically appraise the work of others (AO3) and will test AO1 by requiring you to write extended prose.

Content Guidance

The Content Guidance section is a guide to the content of **Unit 2: The Variety of Living Organisms**. It contains the following features.

Key concepts you must understand

Whereas you can learn facts, these are ideas or concepts that often form the basis of models that we use to explain aspects of biology. You can know the words that describe a concept like the cohesion-tension theory, but you will not be able to use this information unless you really understand what is going on.

STUDENT UNIT GUIDE

NEW EDITION

AQA AS Biology Unit 2
The Variety of Living Organisms

Steve Potter and Martin Rowland

Philip Allan Updates, an imprint of Hodder Education, an Hachette UK company, Market Place, Deddington, Oxfordshire OX15 0SE

Orders
Bookpoint Ltd, 130 Milton Park, Abingdon, Oxfordshire, OX14 4SB
tel: 01235 827827
fax: 01235 400401
e-mail: education@bookpoint.co.uk
Lines are open 9.00 a.m.–5.00 p.m., Monday to Saturday, with a 24-hour message answering service.
You can also order through the Philip Allan Updates website: www.philipallan.co.uk

ISBN 978-1-4441-5289-0

First printed 2011
Impression number 5 4 3 2 1
Year 2015 2014 2013 2012 2011

Cover photo: fusebulb/Fotolia

Printed in Dubai

Hachette UK's policy is to use papers that are natural, renewable and recyclable products and made from wood grown in sustainable forests. The logging and manufacturing processes are expected to conform to the environmental regulations of the country of origin.

P01939

Key facts you must know and understand

These are exactly what you might think: a summary of all the basic knowledge that you must be able to recall and show that you understand. The knowledge has been broken down into a number of small facts that you must learn. This means that the list of 'Key facts' for some topics is quite long. However, this approach makes quite clear *everything* you need to know about the topic.

Summary

This part describes the skills you should be able to demonstrate after studying the relevant topic. These include the skills associated with the assessment objectives that examiners will ask you to demonstrate in the BIOL2 unit test.

Content Guidance

The nature of variation

Key concepts you must understand

The variation within a species is called **intraspecific variation**; that between different species is **interspecific variation**.

Intraspecific variation can be caused by:
- genetic differences
- differences in the environment
- a combination of genetic and environmental influences

A **species** is a group of individuals that are similar physically and physiologically and which can interbreed to produce fertile offspring.

A **population** is the total number of organisms of one species in a given area at a given time.

A **sample** is a subset of a population.

Continuous variation is variation in which there is a continuous range of values, such as in human height. **Discontinuous (categoric) variation** is variation in which only a limited number of discrete (separate) categories are possible, for example, human ABO blood groups.

Continuously variable features are often **normally distributed**. Their variability can be described by the **mean** and the **standard deviation**.

A **gene** is a section of DNA that controls a particular feature (e.g. eye colour). An **allele** is one of two or more versions of a gene (e.g. blue or brown alleles of the gene for eye colour).

examiner tip
Examiners often begin a question by asking candidates to define a term. Ensure you can clearly define the terms species, population, gene and allele.

Key facts you must know and understand

No two organisms are identical; there is always some difference. There are fewer differences between different humans than between humans and other animals. Humans are all members of the same species — *Homo sapiens*. In other words, the extent of intraspecific variation is usually less than the extent of interspecific variation.

Genetic variation can be a result of different genes or different combinations of alleles of the same genes, brought about by:
- mutations
- crossing-over in meiosis (see pages 22–23)
- independent assortment of chromosomes in meiosis (see page 23)
- random fertilisation of gametes (any male gamete could fertilise any available female gamete)

Organisms with the same combination of alleles may still be slightly different due to the effects of the environment. At birth, identical twins are never quite the same; they occupied different positions in the womb and probably had different sized placentas. As a result, one may have received more nutrients than the other and so was born heavier. The flowers of genetically identical hydrangea plants can differ in colour, depending on the pH of the soil (pink in high pH, blue in lower pH).

Tall pea plants are taller than dwarf pea plants because of different alleles of the gene for height, but all tall pea plants are not the same height because of differences in the environment (they may receive different amounts of water, mineral ions, light etc., which affect their growth). So variation in the height of pea plants is due to a combination of genetic and environmental effects. This is illustrated in Figure 1.

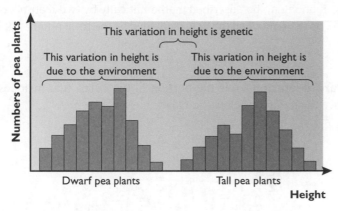

Figure 1 Genetic and environmental variation of height in pea plants

Features showing discontinuous (categoric) variation present an 'either/or' situation — there is a limited number of variants or different categories. For example, blood groups and eye colour show discontinuous variation.

Features showing continuous variation show an uninterrupted range of different types. For example, height in humans, body mass in humans and leaf width in pea plants show continuous variation

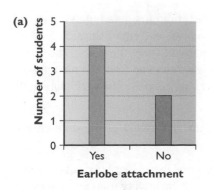

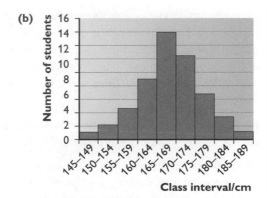

Figure 2 (a) A bar chart of earlobe attachment and (b) a histogram of height in 17-year-old students

Knowledge check I

Figure 2(a) is a bar chart whereas Figure 2(b) is a histogram. Explain the difference between the two and explain why they are used in this way.

The frequencies of each type in a population can be shown in graphs. As Figure 2 shows, we should plot a **bar chart** to show categoric variation but a **histogram** to show continuous variation.

Measuring the variation in a population involves taking a sample of that population — it would be impractical to try to measure every single member of the population. The sample should be a **random sample** to avoid any **bias** in the choosing of individuals. If the sample is large enough, it will probably be quite representative of the whole population, but small samples can easily be unrepresentative.

If enough values are plotted of a feature showing continuous variation, the distribution forms a bell-shaped curve, called a **normal distribution curve**.

A normal distribution can be described mathematically by two features of the curve:
- the **mean** ($\bar{x}$) is the sum of all the values divided by the total number of individuals, i.e.,
$$\bar{x} = \frac{\Sigma x}{n}$$
- the **standard deviation** (σ) is the portion of the curve that includes 34% of all values above the mean or below the mean.

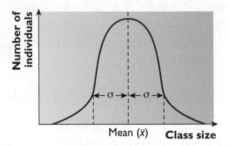

Figure 3 The mean and standard deviation of a typical normal distribution

The mean is a measure of 'central tendency' — it tells you the 'typical value'. The standard deviation is a measure of 'dispersion' — it tells you the extent to which the data are spread about the mean.

The standard deviation is more useful in this respect than just looking at the overall **range** of values. The range is defined solely by the two extreme values; it does not matter where the intervening values lie or how many of them there are. The values at the extremes can easily be 'freakish' — a long way removed from the next nearest values. We call these **outliers** and they distort the impression of the variability of the feature.

However, the standard deviation takes into account the extent of the difference of each and every value from the mean.

The two normal distributions in Figure 4 have the same mean, but have very different standard deviations. Population (a) shows much more variability than population (b), and this is reflected in their standard deviations.

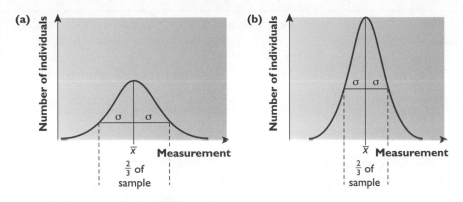

Figure 4 Standard deviation

The tentative nature of experimental results

When presenting the results of their research, scientists try to show the degree of uncertainty in their data. One way of doing this is to repeat their experiments many times and calculate the mean and standard deviation of these results. The standard deviation can be used to calculate another value called the standard error, and the value of the standard error is often shown on graphs as 'error bars' to illustrate the range of results for a given condition. If error bars for two different conditions overlap, it suggests the differences between the two means are probably due to chance. If the error bars don't overlap, then the differences are probably not due to chance. But even then, it is not certain.

Figure 5 shows the effect of two diets on the concentration of glycogen in the muscles of men and women.

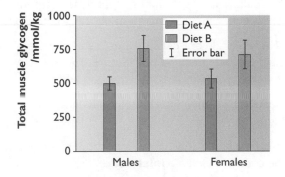

Figure 5 Concentration of glycogen in the muscles of men and women

The error bars show that, in females, the difference between the two diets is probably due to chance, but that this is not the case in males. The error bars also show that, for each diet, the difference between males and females is probably due to chance.

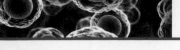
Summary

After studying this topic, you should be able to:

- explain that variation between members of a species (*intra*specific variation) can result from genetic factors, environmental factors or a combination of both

- explain the need for random sampling

- show understanding of how chance events contribute to differences between samples

- show understanding of the concept of a normal distribution about a mean

- explain why standard deviation is a better measure of the spread of data about a mean than is the range of values

- analyse and interpret data relating to intraspecific and interspecific variation

- show an appreciation of the tentative nature of any conclusions that can be drawn about the causes of variation

DNA — its structure and how it works

Key concepts you must understand

DNA is the molecule of inheritance. It is a huge molecule made from two strands wound into a double helix.

One of these strands is called the **coding strand**. This strand is used as a code for specific features. The other strand is not used in this way and is called the **non-coding strand**.

DNA is a stable molecule. This is essential so that genetic information passed on in cell divisions is consistent from one generation of cells to the next.

Sections of the DNA molecule are called genes; each gene is a part of the coding strand and controls the development of one feature. The position of each gene is the **locus** of that gene.

A gene controls a feature by coding for a protein that directly or indirectly determines that feature (see Figure 6).

The order of organic bases in a gene codes for proteins. Since three bases code for one amino acid, the code is called a **triplet code**.

Because of its size, DNA cannot leave the nucleus. Instead it produces 'messenger' molecules that direct the cell to synthesise proteins; these molecules are called **messenger RNA (mRNA)**.

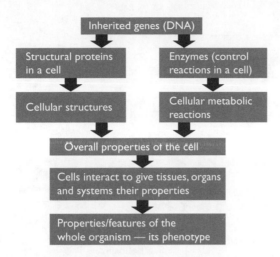

Figure 6 Genes affect the whole organism

DNA contains non-coding regions, both within genes (called **introns**) and between genes (called **minisatellites** and **microsatellites**).

Key facts you must know and understand

The structure of DNA

DNA is made of two polynucleotide strands wound into a double helix. Each polynucleotide strand is made from many structures called nucleotides, held together by covalent bonds.

Look at the nucleotide at the bottom left of Figure 7(a). It is made up of:
- a phosphate group
- a molecule of **deoxyribose** (a pentose sugar — a sugar with five carbon atoms)
- an organic, nitrogenous base (either **adenine**, **thymine**, **cytosine** or **guanine**)

Figure 7(a) also shows how the two strands are arranged in a precise manner.

A nucleotide containing the base adenine on one of the strands is *always* paired with one containing thymine on the other strand. A nucleotide containing the base cytosine is always paired with one containing guanine. This is the **base-pairing rule**. Adenine and thymine are said to be **complementary bases**, as are cytosine and guanine.

The strands are oriented opposite to each other. The 'start' or 'top' of one strand is paired with the 'end' or 'bottom' of the other. The two strands are said to be **anti-parallel**.

examiner tip

You might be given a drawing of part of a DNA molecule and asked to draw a box around one single nucleotide. Make sure your box includes one base, one phosphate group and the molecule of deoxyribose to which they are both attached.

Knowledge check 4

The number of cytosine bases in a molecule of DNA is the same as the number of guanine bases. Explain why.

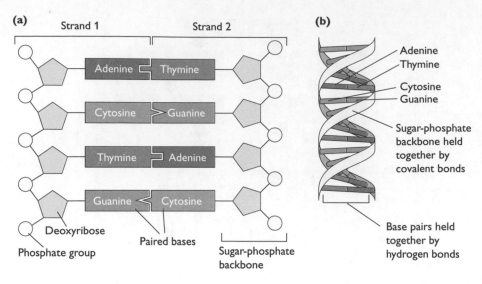

Figure 7 (a) DNA comprises base pairs attached to a sugar-phosphate backbone and (b) formed into a double helix

The covalent bonds holding the nucleotides together in each strand are stronger than the hydrogen bonds holding the two strands together. The strong covalent bonds ensure that the structure of each strand is stable. The hydrogen bonds are much weaker and this allows the two strands to be separated easily. This is important in replication of the DNA (see page 19), in DNA hybridisation (see page 54) and in protein synthesis (which you will study in Unit 5).

Size, location and organisation of DNA in eukaryotic and prokaryotic cells

Feature of DNA	Eukaryotic cell	Prokaryotic cell
Size of molecule	10^8–10^9 base pairs (100–1000 megabase pairs)	10^5–10^6 base pairs (0.1–1 megabase pairs)
Nature of molecule	Linear Associated with histone proteins to form **chromosomes**	Circular Naked (not associated with histone proteins) Some DNA exists as very small, circular **plasmids**
Location of molecule	In nucleus	Free in cytoplasm

The association of eukaryotic DNA with histones, shown in Figure 8, helps to compact the molecule into a smaller space, to protect the DNA and to improve its stability.

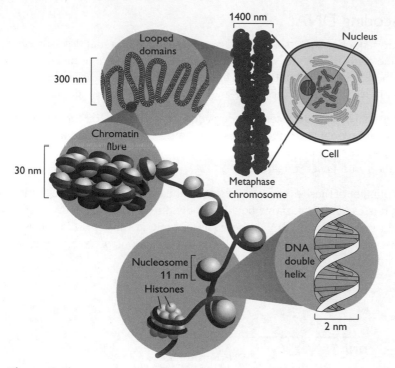

Figure 8 The organisation of DNA into chromosomes in a eukaryotic cell

The base sequence of DNA determines the structure of proteins

Each protein is made from one or more polypeptide chains. A polypeptide contains many amino acids linked together by peptide bonds. The sequence of the amino acids in a polypeptide chain is determined by sequence of the bases in the gene that directs the synthesis of that polypeptide. Each triplet of bases codes for one amino acid.

When a gene is being used to direct the production of its encoded polypeptide, its base sequence is copied to a smaller molecule, called messenger ribonucleic acid (mRNA). Unlike DNA, this molecule can leave the nucleus via a pore in the nuclear envelope, as shown in Figure 9. Its code is used by ribosomes to make polypeptide chains with the encoded sequence of amino acids.

Knowledge check 5

One gene is a different size from another. How is this possible?

Knowledge check 6

Explain why mRNA can leave the nucleus whereas DNA cannot.

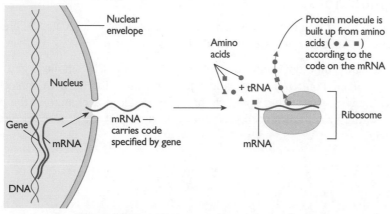

Figure 9 How DNA controls protein synthesis

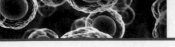

Non-coding DNA

There are sequences of bases within a gene called **introns** that do not code for any of the amino acids in the final polypeptide chain. The coding regions are called **exons**. Figure 10 shows that, when the DNA base sequence of a gene is copied to mRNA, the introns are removed from the mRNA molecule before it leaves the nucleus.

Knowledge check 7

What is a trinucleotide?

There are also base sequences between genes that do not code for amino acids. They often contain repeating trinucleotides containing cytosine and guanine, for example CAG or CGG. These multiple repeats are called minisatellites or microsatellites (which are smaller, with fewer repeats).

A great number of these microsatellites and minisatellites occur in the human genome and each is extremely variable between individuals. The variability of minisatellite DNA and microsatellite DNA is the basis of genetic fingerprinting. It is extremely unlikely that two individuals will have the same versions of several minisatellites or microsatellites.

Knowledge check 8

A particular gene is 450 base pairs long. (a) What is the maximum number of amino acids in the polypeptide which this gene encodes? (b) Why might the actual number be less than this maximum value?

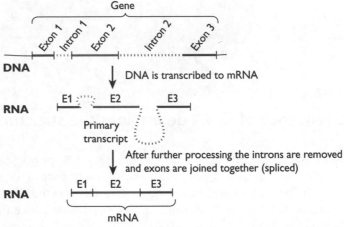

In a gene there are non-coding sections of DNA called introns, separating the coding regions or exons. Introns are transcribed but are later 'cut out' of the final mRNA molecule.

Figure 10 Introns and exons

Gene mutations

A gene mutation is a change in the base sequence of a DNA molecule. Different base sequences of the same gene are called **alleles** of that gene.

There are many different kinds of mutation, but even the smallest can have a significant effect. Point mutations affect just one base in the DNA molecule. As Figure 11 shows, even such a small change may result in a change in the primary structure of the protein produced, as it changes the triplet that codes for one of the amino acids.

Original code	AAT	GCC	TAG	AGG
Amino acids	Leucine	Arginine	Isoleucine	Serine

Changed code	AAG	GCC	TAG	AGG
Amino acids	Phenylalanine	Arginine	Isoleucine	Serine
	CHANGED SEQUENCE			

Figure 11 A change in base sequence may result in a new amino acid

A new amino acid will result in a different polypeptide being synthesised. This could result in either:

- a different structural protein, for example a different haemoglobin, such as that which causes sickle-cell anaemia, or
- a non-functional enzyme, which could result in a metabolic pathway being interrupted

After studying this topic, you should be able to:

- recognise and identify the components of a DNA molecule
- describe how the features of a DNA molecule enable it to act as a stable, information-carrying molecule
- define the term gene and show understanding of how the base-triplet code of a gene determines the sequence of amino acids in a polypeptide chain
- explain that different base sequences of a gene result in alleles of that gene that might result in the formation of non-functional proteins, including non-functional enzymes
- explain that introns and multiple repeats are common non-coding regions of a gene
- describe the differences between DNA molecules in prokaryotes and eukaryotes

Summary

Passing on the genetic material

Key concepts you must understand

DNA is passed on to:
- the next generation of cells (**daughter cells**), when cells in the body divide to bring about growth or repair
- the next generation of individuals

When passing on DNA to daughter cells, the DNA in the **parent cells** is **replicated** precisely so that the daughter cells are genetically identical to each other and to the parent cell.

In most cases, it is advantageous for the next generation of individuals *not* to be genetically identical, but to show genetic variation. This gives the species as a whole a better chance of survival if the environment changes.

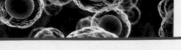
The genetic differences between the **gametes** (sex cells) that fuse to start the next generation produce genetic variation in the next generation of individuals.

Gametes have only half the amount of DNA (and so, half the number of chromosomes) of the parent cell. When two gametes fuse, the full amount of DNA is restored.

The full number of chromosomes is the **diploid number** and the number present in gametes is the **haploid number**.

The chromosomes in human cells exist in pairs called **homologous chromosomes**. By the time a cell enters cell division, each chromosome has been replicated to form a pair of **sister chromatids**.

The chromosomes in each homologous pair carry genes for the same feature in the same sequence, but the alleles of those genes may be different. Figure 12 shows, however, how the alleles on the sister chromatids of one chromosome are identical.

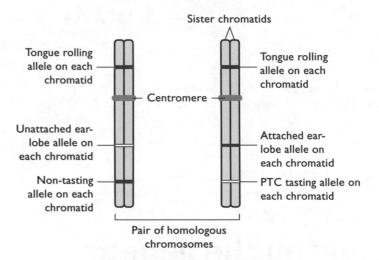

Figure 12 Homologous chromosomes

Knowledge check 9

A human white blood cell has 46 chromosomes. How many homologous pairs does it contain?

When diploid cells divide by mitosis, they produce diploid, genetically identical daughter cells. When parent cells divide by meiosis, they produce haploid, genetically different daughter cells.

Cell division by mitosis normally takes place in a very controlled way. It is regulated by a number of genes. However, sometimes these controls fail and mitosis takes place in an uncontrolled manner. This leads to the formation of a **tumour**.

Tumours can be **benign** or **malignant**. Benign tumours are usually slow-growing and do not spread to other parts of the body. Malignant tumours grow much more quickly and often spread. It is these tumours that form **cancers**.

Key facts you must know and understand

DNA replication

DNA molecules exist within chromosomes in the nucleus and are surrounded by a 'soup' of free DNA nucleotides. These nucleotides are used to build new strands of DNA. The process involves the two enzymes shown in Figure 13(a), but the key stages are as follows.

- Molecules of the enzyme **DNA helicase** break hydrogen bonds holding the two polynucleotide strands together and 'unwind' part of the DNA molecule, revealing two single-stranded regions.
- Molecules of **DNA polymerase** follow the helicase along each single-stranded region, which acts as a template for the synthesis of a new strand.
- The DNA polymerase assembles free DNA nucleotides into a new strand alongside each of the template strands. The base sequence in each of these new strands is complementary to its template strand because of the base-pairing rule — A-T, C-G (see page 13).
- The processes of unwinding followed by complementary strand synthesis progress along the whole length of the DNA molecule.

> **Knowledge check 10**
>
> If all the free DNA nucleotides used in DNA replication were labelled, what proportion of labelled nucleotides would be present in each new DNA molecule?

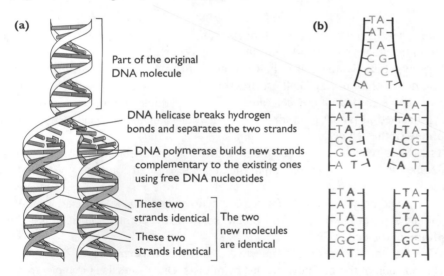

Figure 13 (a) DNA replication; (b) semi-conservative replication of DNA

This method of replicating DNA is called **semi-conservative replication**. Figure 13(b) shows that:

- each new DNA molecule contains one strand from the original DNA and one newly synthesised strand
- both new DNA molecules are identical to each other and to the original molecule

The cell cycle

Cells that are able to divide go through a cycle of events called the **cell cycle**, shown in Figure 14. These events ultimately allow the cell to divide to form two cells by mitotic division.

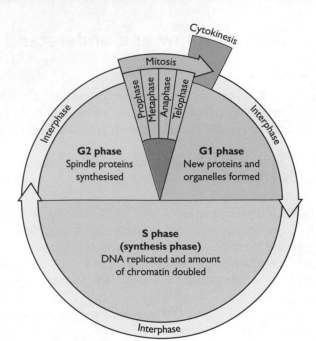

Figure 14 The cell cycle

Before it can divide again, a cell that has just been produced by cell division goes through the stages shown in Figure 14.

- It grows. Initially, the cell is half the size of the parent cell, with only half the organelles of a full-sized cell. During this phase of the cell cycle, more organelles are synthesised and the cell enlarges. Nucleotides and histone proteins are synthesised in preparation for DNA replication later in the cycle. This is the **G1 phase** of the cell cycle.
- Its DNA is replicated and combines with newly synthesised histone proteins to double the amount of chromatin in the nucleus. The cell continues to grow. This is the **S phase** of the cell cycle.
- Specialised proteins called tubulins are synthesised. These are used to make the spindle apparatus, which will eventually separate the chromosomes. This is the **G2 phase** of the cell cycle.
- The G1, S and G2 phases are collectively known as **interphase**.
- The nucleus of the cell now divides by mitosis. Once mitosis is complete, the cell divides into two cells during **cytokinesis**.

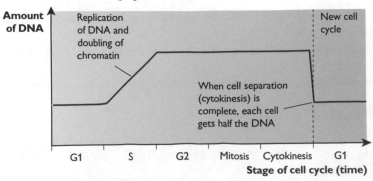

Figure 15 The changes in DNA content during the cell cycle

owledge check 11

in why the fall in the
t of DNA shown in
5 occurs so late in
's.

Mitosis

Mitosis involves the division of the nucleus. In most cases, this is followed by the cytoplasm dividing and producing two cells (cytokinesis).

Mitosis results in two nuclei with the same number and type of chromosomes as each other and as the parent cell that formed them. The cells containing these nuclei are genetically identical.

Mitosis is divided into the four key stages shown in Figure 16: **prophase**, **metaphase**, **anaphase** and **telophase**.

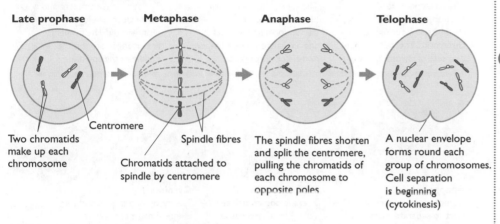

Late prophase

Two chromatids make up each chromosome

Centromere

Metaphase

Spindle fibres

Chromatids attached to spindle by centromere

Anaphase

The spindle fibres shorten and split the centromere, pulling the chromatids of each chromosome to opposite poles

Telophase

A nuclear envelope forms round each group of chromosomes. Cell separation is beginning (cytokinesis)

Figure 16 Stages of mitosis

Stage of mitosis	Main events
Prophase	Each chromosome coils and becomes visible as two chromatids held together by a **centromere**. (During interphase, the DNA was replicated and from one chromosome two identical **sister chromatids** were formed.) The nuclear envelope starts to break down.
Metaphase	The spindle forms. The centromeres attach the chromatids to the spindle fibres so that they lie across the middle of the spindle.
Anaphase	The centromeres divide. The spindle fibres shorten and pull the sister chromatids to opposite poles of the cell. Once the chromatids have been separated, they are called chromosomes again.
Telophase	The spindle fibres are broken down. The two sets of chromosomes group together at each pole and a nuclear envelope forms around each. The chromosomes uncoil and cannot be seen as individual structures.

Meiosis

A meiotic cell division involves two divisions of the nucleus (**meiosis**) and two divisions of the cell (**cytokinesis**).

- Meiosis I separates the chromosomes from each homologous pair into different cells, halving the chromosome number.
- Meiosis II separates the chromatids in each chromosome, rather like mitosis.

Knowledge check 12

The type of cell division that occurs in prokaryotes is not referred to as mitosis. Explain why.

examiner tip

Make sure you recognise the main features of each stage of mitosis in drawings and in photographs.

Knowledge check 13

Cancers are caused by cells that divide in an uncontrolled way. Vincristine is a drug that prevents spindle formation in cells. Explain why vincristine can be used to treat cancer.

examiner tip

Be quite clear in your mind that the *chromosome number* is halved after the first meiotic division; the cells formed at this stage are haploid cells.

The main stages of meiosis I

These are outlined in Figure 17.

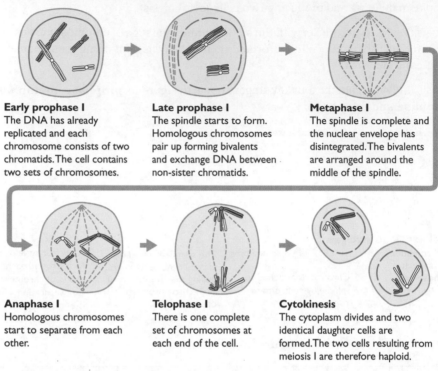

Early prophase I
The DNA has already replicated and each chromosome consists of two chromatids. The cell contains two sets of chromosomes.

Late prophase I
The spindle starts to form. Homologous chromosomes pair up forming bivalents and exchange DNA between non-sister chromatids.

Metaphase I
The spindle is complete and the nuclear envelope has disintegrated. The bivalents are arranged around the middle of the spindle.

Anaphase I
Homologous chromosomes start to separate from each other.

Telophase I
There is one complete set of chromosomes at each end of the cell.

Cytokinesis
The cytoplasm divides and two identical daughter cells are formed. The two cells resulting from meiosis I are therefore haploid.

Figure 17 Stages of meiosis I

Crossing over

Crossing over is a 'cut-and-paste' event that occurs during prophase 1. Figure 18 shows how this produces new combinations of alleles from maternal and paternal chromosomes.

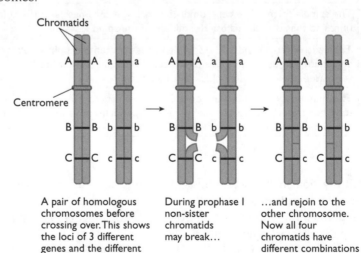

A pair of homologous chromosomes before crossing over. This shows the loci of 3 different genes and the different alleles of these genes on the chromosomes

During prophase I non-sister chromatids may break...

...and rejoin to the other chromosome. Now all four chromatids have different combinations of alleles — more variation exists

Figure 18 Crossing over produces new combinations of alleles

Knowledge check 14

After a single cross-over, how many chromatids have a new combination of alleles?

Random segregation

Figure 19 shows that the way homologous chromosomes are aligned at metaphase determines how they will be segregated into the two new cells. The letters represent the alleles of genes A to E.

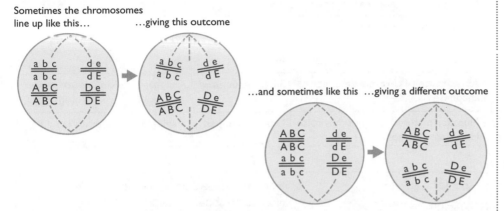

Figure 19 Random segregation of homologous chromosomes gives rise to genetic variation

The main stages of meiosis II

These are outlined in Figure 20.

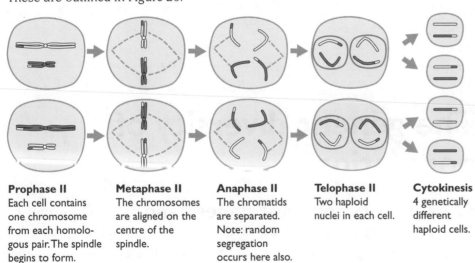

Prophase II
Each cell contains one chromosome from each homologous pair. The spindle begins to form.

Metaphase II
The chromosomes are aligned on the centre of the spindle.

Anaphase II
The chromatids are separated. Note: random segregation occurs here also.

Telophase II
Two haploid nuclei in each cell.

Cytokinesis
4 genetically different haploid cells.

Figure 20 Stages of meiosis II

Meiotic cell division

The key features of a meiotic cell division are that:
- it involves two nuclear divisions
- it forms four daughter cells
- the daughter cells are haploid
- the daughter cells show genetic variation (contain the same genes, but different combinations of alleles)

examiner tip
You can use the number and appearance of the chromosomes at anaphase to identify a cell division as mitotic or meiotic. If the chromosomes are double structures (two chromatids) it can only be anaphase I of meiosis. If the chromosomes are single structures, it could be mitosis or anaphase II of meiosis. In that case, look at how many there are. If the diploid number is moving to each pole, it must be mitosis; if the haploid number is moving to each pole, it must be meiosis (II).

Meiosis and mitosis compared

Feature of the process	Mitosis	Meiosis
Number of nuclear divisions	One	Two
Number of daughter cells formed	Two	Four
Chromosome number of daughter cells	Diploid	Haploid
Genetic variation in daughter cells	No	Yes
Appearance of chromosomes at anaphase	Single structures	Anaphase I — double structures Anaphase II — single structures
Number of chromosomes moving to each pole at anaphase	Diploid number	Haploid number (both I and II)

Summary

After studying this topic, you should be able to:

- describe the process of semi-conservative replication of DNA, including the role of enzymes
- recognise, from drawings or photographs, and name the stages of mitosis
- explain the events that occur during each stage of mitosis
- explain that mitosis produces cells that are genetically identical, i.e. clones
- interpret data relating to the cell cycle
- relate your understanding of the cell cycle to information about cancer and its treatment
- recognise that meiosis produces cells that are haploid
- show understanding that meiosis produces cells that are genetically different as a result of crossing over between homologous chromosomes and the independent segregation of homologous chromosomes

The genetic diversity of populations

Key concepts you must understand

Mutations create new alleles of genes. Crossing over and random segregation of chromosomes during meiosis, together with random fertilisation of gametes during sexual reproduction, create new combinations of alleles.

Random fertilisation results in even more genetic diversity if mating is also random. This type of mating is called **out-breeding**.

The **founder effect**, **genetic bottlenecks**, **in-breeding** and **selective breeding** reduce genetic diversity.

Key facts you must know and understand

The founder effect

In the founder effect, a few individuals colonise a new environment and breed there.

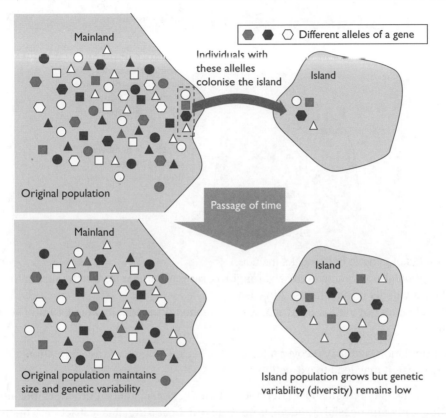

Figure 21 Consequences of the founder effect

Only a fraction of the alleles that were present in the original population is present in the few individuals that colonise the new environment and establish a new population. Therefore, the genetic diversity of the new population will be reduced (Figure 21).

Examples of the founder effect include:
- the Afrikaner population of South Africa, which is descended from just a few Dutch settlers
- the grey squirrel population in the UK, which is descended from a small number of colonising individuals

Genetic bottlenecks

In a genetic bottleneck, the population is reduced drastically by some environmental effect, such as a disease. Figure 22 shows that only a fraction of the alleles that were present in the original population is present in the small number of individuals that survive and breed. Therefore, the genetic diversity of the new population will be reduced.

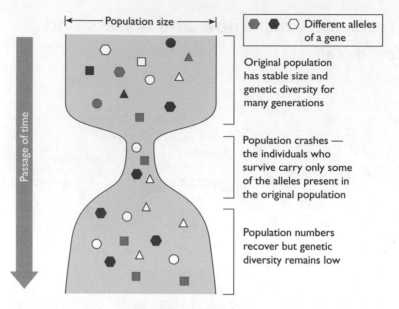

Figure 22 A genetic bottleneck

Examples of genetic bottlenecks include:
- the population of lions in the Ngorongoro Crater in Tanzania, which crashed from 75 individuals in 1962 to just 7 in 1964
- the cheetah population in Africa, which crashed dramatically about 10 000 years ago

The population size may recover after a bottleneck event, but if the population crash is too great, it may lead to extinction.

Genetic diversity remains low after a founder effect or a genetic bottleneck because mating, although possibly random, is not true **out-breeding**. It is restricted to mating between the few genetic types that survive in the new population and is effectively an example of **in-breeding**, where mating is restricted to just a few types. In-breeding maintains a low genetic diversity.

Conservation projects designed to maintain and increase populations of threatened species (such as the cheetah populations) often encourage breeding between animals from different populations. This is a kind of forced out-breeding designed to maximise genetic diversity in each population by introducing alleles from other populations.

Selective breeding

Selective breeding is a form of in-breeding that is controlled by humans. For thousands of years, genetic modification for increased yields was achieved by careful breeding of selected strains of stock animals and crop plants. This meant only allowing breeding between high-yielding plants or animals, with the result that many of the alleles in the original gene pool were excluded from subsequent generations. They were only passed on if, by chance, they were found in an organism that had desirable alleles for, say, increased milk yield.

Examples of selective breeding include:
- cattle bred for high meat or milk yield
- pigs, descended from wild boars and bred to produce large litters (between 12 and 20 piglets) that grow quickly
- modern wheat, descended from grasses and selectively bred to give a high grain yield and be resistant to diseases

After studying this topic, you should be able to:
- explain that genetic diversity results from differences in DNA
- explain how genetic diversity is increased by gene mutation, crossing over, random segregation of chromosomes, random fertilisation and out-breeding

- explain how genetic diversity is decreased by the founder effect, genetic bottlenecks, in-breeding and artificial selection
- discuss ethical issues involved in the selection of domesticated animals

Summary

Selection in a population

Key concepts you must understand

As a result of his observations, Charles Darwin concluded that:
- all species tend to produce more offspring than can possibly survive
- there is variation among the offspring

From these observations he deduced that:
- there will be a 'struggle for existence' between members of a species, because they over-reproduce, and resources are limited
- because of variation, some members of a species will be better adapted than others to their environment

Combining these two deductions, Darwin proposed:

> Those members of a species which are best adapted to their environment will survive and reproduce in greater numbers than others less well adapted.

This was his now famous theory of **natural selection**.

Darwin knew little of genetics. However, we can modify his theory to take account of gene action. Genes or, more precisely, alleles of genes, determine features. Suppose an allele determines a feature that gives an organism an advantage in its environment. As the individuals with the advantage survive and reproduce in greater numbers, the frequency of the advantageous allele in the population will increase.

Mutations introduce new alleles into populations. Any mutation could:
- confer a selective advantage — the frequency of the allele will increase over time
- be neutral in its overall effect — the frequency may increase slowly, remain stable or decrease (the change in frequency will depend on what other genes/alleles are associated with the mutation)
- be disadvantageous — the frequency of the allele will be low and the allele could disappear from the population

examiner tip
Do not write that selection changes gene frequencies. Selection — both artificial and natural — changes the frequency of the alleles of a gene.

Key facts you must know and understand

The effects of antibiotics on bacterial populations

Antibiotics act against bacteria by disrupting cellular processes such as:

- formation of cell walls
- DNA replication
- protein synthesis

Some antibiotics kill bacteria — these are **bactericidal** antibiotics. Others do not kill bacteria but stop them reproducing — these are **bacteriostatic** antibiotics.

The ways in which some different antibiotics act are summarised in the table below.

Mode of action of antibiotic	Example	How the antibiotic works	Bactericidal or bacteriostatic?
Prevents formation of cell wall	Penicillin	Weakened cell wall cannot resist entry of water by osmosis and cell bursts (osmotic lysis)	Bactericidal
Disrupts DNA replication	Ciprofloxacin	Bacteria are not killed, but cell division is halted	Bacteriostatic
Disrupts protein synthesis	Tetracycline	Bacterial cell cannot synthesise enzymes and structural proteins	Bactericidal

Antibiotics that disrupt cell wall synthesis interfere with the synthesis of the peptidoglycan layer in the cell wall. Ordinarily, osmosis into a bacterial cell is resisted by its cell wall. With only an incomplete cell wall to resist the swelling caused by the entry of water, the bacterial cell bursts.

How antibiotic-resistant populations of bacteria develop

Bacterial cells contain two types of DNA. Most bacterial DNA is organised into a single large, cyclical molecule, but some bacterial DNA is found as **plasmids**. These are small, circular molecules of DNA, separate from the main DNA.

Mutations in the plasmid DNA can produce resistance to an antibiotic. Bacteria can pass on these mutations in two ways:

- **Vertical gene transmission**. Bacteria reproduce by a process called **binary fission** (Figure 23). All the DNA in the bacterium replicates prior to reproduction, not just the main DNA molecule. When the bacterial cell divides, each daughter cell will receive some plasmids that carry the gene for resistance to the antibiotic.
- **Horizontal gene transmission**. Some bacteria can also pass plasmids to other bacteria that happen to be around — as well as receive plasmids from these bacteria. The process is a kind of genetic swap-shop called **conjugation**. Plasmids pass through **conjugation tubes** from one bacterium to another (Figure 24).

Bacterial cell about to divide by binary fission

Main DNA molecule

Bacterial cells formed from division

Main DNA molecule

Plasmid carrying gene for antibiotic resistance

Plasmid carrying gene for antibiotic resistance

Both 'daughter' bacteria are resistant to this antibiotic

Figure 23 The gene for antibiotic resistance is passed on when bacteria reproduce by binary fission

examiner tip
Mitosis only occurs in cells with a nucleus, i.e. eukaryotic cells. Do not tell examiners that bacteria reproduce by mitosis.

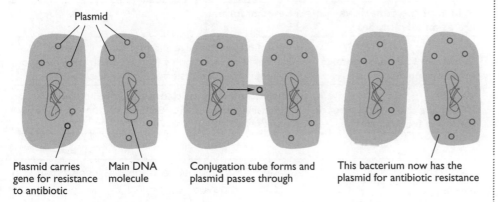

Plasmid

Plasmid carries gene for resistance to antibiotic

Main DNA molecule

Conjugation tube forms and plasmid passes through

This bacterium now has the plasmid for antibiotic resistance

Figure 24 How plasmids can pass from one bacterium to another

Figure 25 shows how conjugation can lead to some bacteria becoming resistant to *several* antibiotics — multiple antibiotic resistance.

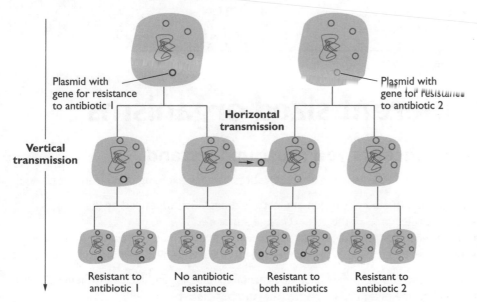

Plasmid with gene for resistance to antibiotic 1

Plasmid with gene for resistance to antibiotic 2

Horizontal transmission

Vertical transmission

Resistant to antibiotic 1

No antibiotic resistance

Resistant to both antibiotics

Resistant to antibiotic 2

Figure 25 The importance of vertical and horizontal transmission in passing on antibiotic resistance in bacteria

Bacteria trapped in polar ice-caps thousands of years ago have been found to contain genes for antibiotic resistance. How is this possible?

examiner tip

You can use the left-hand column of this table to structure your explanation of any novel information in an exam question about natural selection. Just ensure you do *not* suggest that the factor creating the selection pressure *causes* the relevant mutation.

A mutation conferring resistance only gives a selective advantage if the antibiotic is being used. In this case, it creates a **selection pressure** in favour of those bacteria that have the allele for resistance and against those that do not have it.

More of the bacteria with the allele for resistance survive to reproduce than those without it. Thus, the frequency of the allele for resistance increases and will increase with each succeeding generation of bacteria, until nearly all the population carries the resistant allele. This is natural selection and is summarised in the table below.

What factor creates the selection pressure?	Presence of an antibiotic
How did the existing genetic variation arise?	Chance mutation in some individuals confers resistance
Which of the genetic variants are at an advantage?	Resistant forms
What are the consequences for each phenotype?	Resistant forms survive and reproduce in greater numbers than susceptible forms — with time, more of the population are resistant
What are the consequences for allele frequencies?	Alleles conferring resistance are passed on in increasing numbers with each generation — frequency increases

Summary

After studying this topic, you should be able to:

- describe two ways in which antibiotics prevent the growth of bacteria
- explain that resistance to antibiotics can result from mutation of genes, usually carried on plasmids within bacterial cells
- describe how genes for resistance to antibiotics can be passed within a species by vertical gene transmission and between species by horizontal gene transmission
- use the terms adaptation and selection to show how the frequency of an allele conferring resistance to antibiotics can increase in a bacterial population
- apply the concepts of adaptation and selection to other examples

Different sized organisms

Key concepts you must understand

The size of an organism has big implications for gas exchange and transport.

The rate of diffusion of oxygen and carbon dioxide is influenced by the:
- total surface area of the gas exchange surface
- difference in concentration across the exchange surface (diffusion gradient)
- thickness of the exchange surface (length of the diffusion pathway)

The rate at which oxygen is used and carbon dioxide is produced is influenced by the:
- total volume of an organism (i.e. the number of respiring cells)

- metabolic rate (the rate at which energy-consuming processes are happening)
- activity and temperature (which influence metabolic rate)

The ratio between surface area and volume is a crude measure of the ratio of supply of oxygen to the demand for oxygen.

If the organism is a cube and the length of each edge is 1 arbitrary unit (au), then:
- the area of each face is $1 \times 1 = 1$ au^2
- there are six faces to a cube, so the total surface area is 6 au^2
- the volume is $1 \times 1 \times 1 = 1$ au^3
- the ratio of surface area to volume = 6/1 = 6

In a bigger organism, where the length of each edge is 2 au:
- the area of each face is $2 \times 2 = 4$ au^2
- there are six faces to a cube, so the total surface area is 24 au^2
- the volume is $2 \times 2 \times 2 = 8$ au^3
- the ratio of surface area to volume = 24/8 = 3

If you calculate the values for cubes of edge 4 au and 8 au, you will find that the surface area-to-volume ratios are 1.5 and 0.75 respectively.

Figure 26 summarises the effect of increasing body size on surface area to volume ratio.

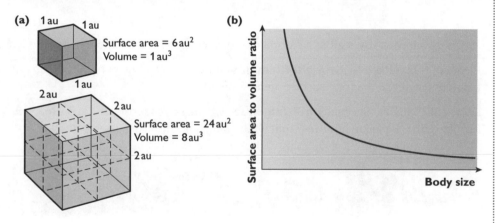

Figure 26 (a) The effect of increasing size on the surface area and volume of a cube (b) The relationship between size and surface area to volume ratio

Very small organisms obtain oxygen by simple diffusion through their surface. The demand for oxygen is satisfied by the rate at which it can be supplied.

In bigger organisms, simple diffusion is too slow to distribute substances around the body. Bigger organisms have:
- specialised gas exchange surfaces
- mechanisms to ventilate their gas exchange surfaces
- mass transport systems

examiner tip
Ensure you write about the surface area-to-volume ratio and not about surface area. Candidates commonly tell examiners that a mouse has a bigger surface area than an elephant, which clearly is not the case. A mouse does have a bigger surface area-to-volume ratio, though.

Knowledge check 19
How does ventilating its gas exchange surface increase an animal's gas exchange rate?

Key facts you must know and understand

Gas exchange surfaces in different organisms

Unicellular protoctists

A single-celled organism, such as the *Amoeba* in Figure 27, exchanges gases through its plasma membrane, which is also its body surface. The large surface area-to-volume ratio means that demand for oxygen is unlikely to outstrip supply.

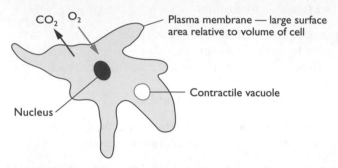

Figure 27 Gas exchange in *Amoeba* (a unicellular protoctist)

Fish

Figure 28 shows a fish's **gills**, which extract oxygen dissolved in water.

Each gill is divided into many **gill filaments** attached to a bony gill arch. These gill filaments increase the total surface area of the gill.

Each gill filament has many **gill lamellae**, which increase the surface area still further. These lamellae have very thin walls that shorten the diffusion pathway.

Knowledge check 20

Air contains more oxygen than water yet a fish dies from lack of oxygen if removed from the water. Use your knowledge of surface area-to-volume ratio to suggest why.

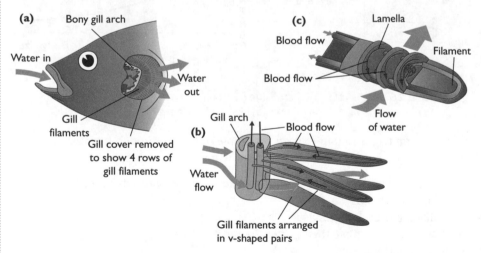

Figure 28 (a) The position of gills (b) the structure of gill filaments (c) blood flow in lamellae

Notice in Figure 28(c) that water flows past the lamellae in the opposite direction to the blood within the lamellae. This **counter current** maintains a concentration gradient for both carbon dioxide and oxygen between the water and the blood, and improves the efficiency of gas exchange.

Insects

Insects do not have lungs or gills, but a **tracheal system** with two main **tracheae** running the length of the insect's body. Figure 29 shows how each trachea opens to the air through several **spiracles** along its length.

Smaller tubes, called **tracheoles**, branch off the tracheae and carry air directly into the cells of the insect's body. The large numbers of tracheoles create a large gas exchange surface and the thin wall of the smallest of these ensures a short diffusion distance.

The air sacs function as a temporary store. In hot conditions, some insects close their spiracles and use oxygen from air stored in the sacs, thus reducing water loss by evaporation.

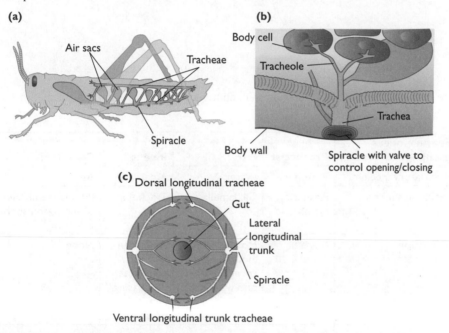

Figure 29 (a) The tracheal system of an insect (b) The relationship between spiracles, tracheae and tracheoles (c) Transverse section of the insect tracheal system

In larger insects there is some ventilation of the system, resulting in mass flow. By opening some spiracles and closing others while dilating and constricting the abdomen, air can be moved along the trachea. However, individual gases, rather than air, still *diffuse* to and from cells along the tracheoles.

The spongy mesophyll of a leaf

Because they are loosely packed, the surface area of spongy mesophyll cells in contact with the air spaces is large enough to allow efficient diffusion of gases. Figure 30 shows that the air spaces also create a free diffusion pathway from stomata to the palisade cells at the top of the leaf.

Cuticle

Upper
epidermis

Palisade
mesophyll

Spongy
mesophyll

Lower
epidermis

Cuticle   Guard cell   Stoma

Daytime
→ CO_2
→ O_2
Night
→ O_2
→ CO_2

Knowledge check 22

Why do the arrows in
Figure 30 represent *net* gas
exchange?

Figure 30 Net gas exchange in a leaf during the day and at night

The principal features of these four exchange surfaces are compared in the table below.

Feature of gas exchange surface	Protoctist	Fish	Insect	Plant leaf
Respiratory medium	Water	Water	Air	Air
Exchange surface	Plasma membrane	Gill lamellae	Tracheoles	Plasma membranes of spongy mesophyll cells
Ventilation	None	Movements of mouth and gill cover create one-way flow	Abdomen dilates/ contracts decreasing/ increasing pressure	None
Large surface area-to-volume ratio due to:	Small volume of cell	Large area of lamellae	Large area of tracheoles	Large area of cell surfaces and loose packing of cells
Oxygen concentration gradient maintained by:	Use of oxygen in cell	Ventilation/ counter-current system in lamellae	Ventilation/ use of oxygen in body cells	Use of oxygen by mesophyll cells
Exchange surface thin due to:	Thin plasma membrane	Thin layer of cells in walls of lamellae	Thin walls of tracheoles	Only cell wall and plasma membrane at exchange surface

Transport systems in different organisms

Over large distances, diffusion becomes an inefficient exchange method. Mammals and plants have evolved efficient systems for mass transport.

Mammals

Blood is moved through a system of blood vessels by the pumping of the heart. Mammals have a double circulation, in which blood passes through the heart twice in a complete circulation of the body. Figure 31 shows the components of this system that you are expected to recall.

There are three main types of blood vessel, which are shown in Figure 32.
- **Arteries** carry blood under high pressure away from the heart to the organs (they branch into smaller arteries called **arterioles**).
- **Veins** carry blood under low pressure away from the organs towards the heart (veins are formed as smaller vessels, called **venules**, fuse).
- **Capillaries** carry blood close to every cell within an organ.

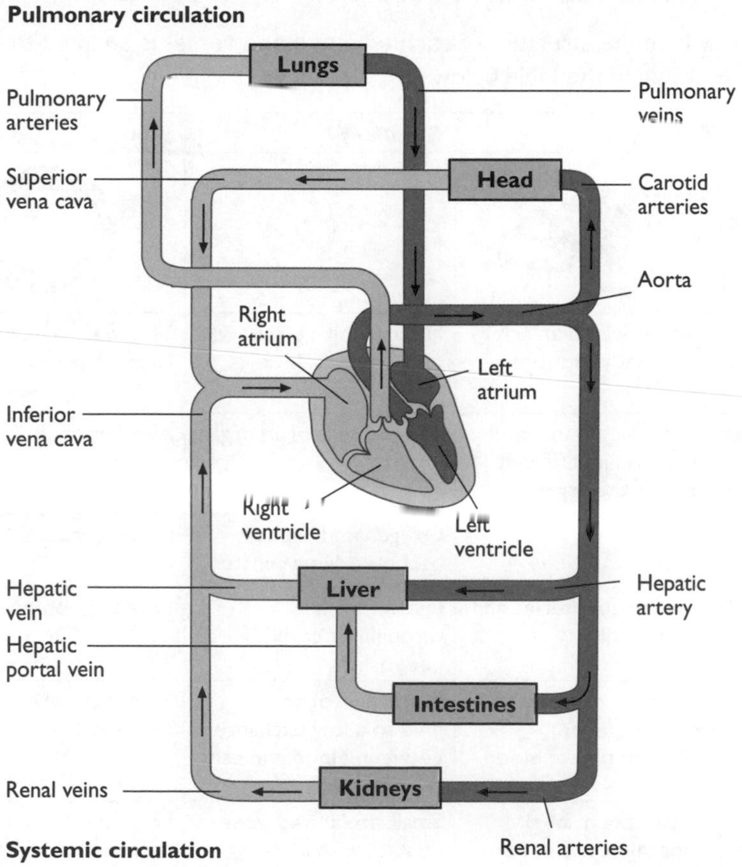

Pulmonary circulation

Figure 31 The main components of the circulatory system of a mammal

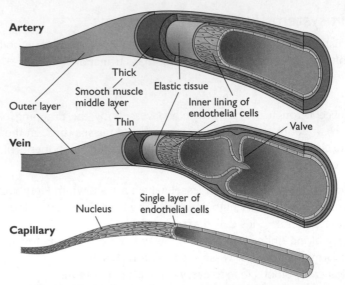

Figure 32 Structure of an artery, a vein and a capillary

Knowledge check 23

How does elastic tissue help to smooth the blood flow in arteries leaving the heart?

The ways in which the structure of each type of blood vessel is adapted to its main function are shown in the table below.

Feature	Artery	Arteriole	Capillary	Vein
Cross-section of vessel				
Structural features	Thick wall and small lumen	Thinner wall than artery with relatively more muscle	Microscopic vessels, wall only one cell thick	Thin wall, little muscle, large lumen, valves
Blood flow	Away from heart, towards an organ	Within an organ, to capillaries in different parts of the organ	Around cells of an organ	Away from an organ towards the heart
Type of blood	Oxygenated*	Oxygenated*	Oxygenated * blood becomes deoxygenated	Deoxygenated*
Blood pressure	High and in pulses (pulsatile)	Lower than arteries and less pulsatile	Pressure falls throughout capillary network	Low and non-pulsatile
Main functions of vessels	Transport of blood to organs	Transport of blood within an organ; redistribution of blood flow	Formation of tissue fluid to allow exchange between blood and cells of an organ	Transport of blood back to the heart
Adaptations to main function	Large amount of elastic tissue allows stretching due to surges in pressure and recoil afterwards; endothelium provides smooth inner surface to reduce resistance	Large amount of smooth muscle under nervous control allows redistribution of blood; constriction limits blood flow, dilation increases blood flow	Small size allows close contact with all cells in the body; thin, permeable (leaky) walls allow formation of tissue fluid for exchange	Large lumen and thin wall offer least resistance to blood flow as blood is under low pressure; valves prevent backflow of blood

* reversed in pulmonary arteries and veins

The formation of tissue fluid and exchange between blood and cells

Blood flows close to every cell of the body in the capillary networks, in all organs. It is tissue fluid, not blood, however, which carries glucose and oxygen to the cells as well as carbon dioxide and other waste products in the opposite direction. Figure 33 shows the circulation of tissue fluid that occurs in every capillary network.

Tissue fluid bathes cells in a fluid that provides a constant environment. The constant pH and temperature of the tissue fluid help to provide optimum conditions for enzyme activity in the cells.

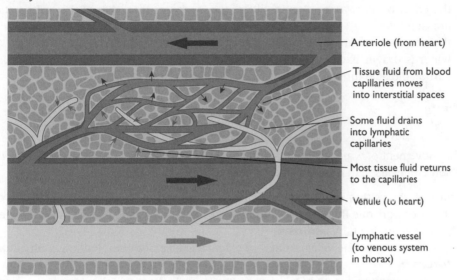

Arteriole (from heart)

Tissue fluid from blood capillaries moves into interstitial spaces

Some fluid drains into lymphatic capillaries

Most tissue fluid returns to the capillaries

Venule (to heart)

Lymphatic vessel (to venous system in thorax)

Figure 33 The circulation of tissue fluid and the formation of lymph

Tissue fluid forms because the capillary walls are permeable to ions and small molecules. However, **plasma protein** molecules are too large to escape and so are not normally found in tissue fluid.

Figure 34 shows that two forces influence the circulation of tissue fluid:

- the hydrostatic pressure of the blood (due to the pumping of the heart), which tends to force ions and small molecules out of the capillaries
- the difference in water potentials between the plasma and the surrounding tissue fluid, which could act either way, depending on the balance

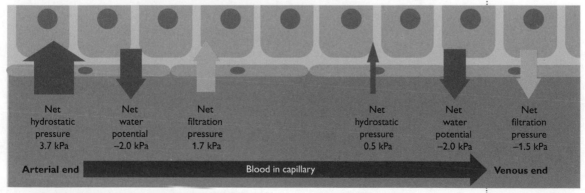

Net hydrostatic pressure 3.7 kPa

Net water potential −2.0 kPa

Net filtration pressure 1.7 kPa

Net hydrostatic pressure 0.5 kPa

Net water potential −2.0 kPa

Net filtration pressure −1.5 kPa

Arterial end　　　Blood in capillary　　　**Venous end**

Figure 34 The forces involved in the formation and reabsorption of tissue fluid

Does a high water potential have a more negative or less negative value than a low water potential?

At the arterial end of a capillary network:
- the hydrostatic effect is greater than that of the water potential difference
- there is a net outward pressure
- tissue fluid (all the substances in the plasma except proteins) is forced out of the capillaries

The loss of fluid reduces the hydrostatic pressure of the blood, whereas the water potential remains more or less unchanged.

Consequently, at the venous end of the capillary network:
- the effect of the water potential difference is greater than that of the hydrostatic pressure
- there is a net inward force due to water potential
- water is drawn back into the capillaries by osmosis; other substances (such as carbon dioxide) diffuse into the blood down concentration gradients

Plants

Transpiration

The movement of water through a plant is called **transpiration**, although this term is sometimes used to describe just the loss of water from the leaves.

Figure 35 shows that water moves through a plant in the following ways.
- It moves from one living cell to another (across the roots and leaves) down a water potential gradient, by **osmosis**.
- It moves through the xylem from root to leaf because of a combination of **physical forces**:
  – **root pressure** — a physical upwards push due to more water entering the xylem vessels

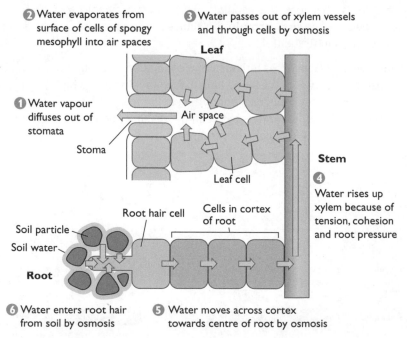

Figure 35 Transpiration — the movement of water through a plant

- **tension** — a state of negative pressure due to evaporation of water from the leaves (a pull)
- **cohesion** — an attractive force between water molecules due to hydrogen bonding
- It **evaporates** from the surfaces of cells in the spongy mesophyll into the air spaces.
- It **diffuses** down a water potential gradient from the air spaces of the spongy mesophyll of the leaf, through open stomata and into the atmosphere.

How does water enter and move across the root?

Water enters the root epidermal cells, particularly the root hair cells (which increase the surface area available for absorption), by osmosis down a water potential gradient. A water potential gradient also exists across the root; the epidermal cells have a higher (less negative) water potential than cells in the centre of the root. Therefore, water moves, by osmosis, through the cortex towards the centre of the root where the xylem is found.

Figure 36 shows the two main pathways by which water moves through the root:
- the **symplast** pathway — water moves through the membranes and cytoplasm of cells
- the **apoplast** pathway — water moves only through cell walls and intercellular spaces

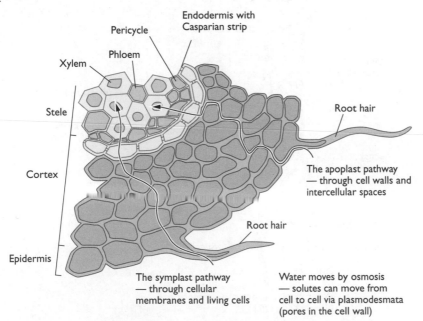

Figure 36 Movement of water across a root

The cells of the endodermis have a layer of suberin (a waxy substance) called the **Casparian strip** in their walls. This acts as an apoplast block, preventing water from moving through the cell walls of the endodermal cells.

Cells surrounding the xylem elements in the root actively secrete ions into the **xylem**, reducing their water potential. Water then follows by osmosis. As more and more

Knowledge check 25

How does the structure of a cell wall allow the passage of water along the apoplast route?

Knowledge check 26

Why would lack of oxygen reduce the effect of root pressure?

water enters the xylem in the centre of the root, it creates a 'root pressure', which forces the water up the xylem.

How does water move up the stem?

Water moves up the stem in the xylem vessels, which form continuous narrow tubes from roots to leaves, because:

- water is lost from the xylem in the leaves, which creates a tension (negative pressure) at the top of the water column in the xylem
- the cohesive force between the water molecules is stronger than the force of the tension

The water molecules effectively form a continuous column, so as the 'top end' is pulled upwards, the rest of the column follows. The root pressure, due to more water entering the xylem in the roots, also helps by giving a 'push'. However, most of the force comes from the tension created by the loss of water from the leaves.

How is water lost from the leaves?

Water is lost from the leaves because a water potential gradient exists from the xylem in the leaf ($\Psi \approx -0.5$ MPa) to the leaf cells ($\Psi \approx -1.5$ MPa) to the air spaces ($\Psi \approx -10$ MPa) and finally to the atmosphere ($\Psi \approx -13$ to -120 MPa). Figure 37 shows that when the guard cells open the stomata, water moves down this water potential gradient.

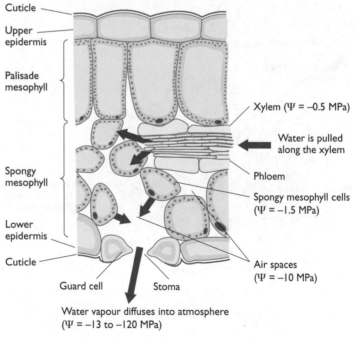

Figure 37 How water moves through leaves

Factors affecting the rate of transpiration

These factors can be grouped into two categories:

- those that affect the water potential gradient between the air spaces in the spongy mesophyll and the atmosphere
- those that affect the total stomatal aperture (effectively, this represents the surface area available for diffusion of water out of the leaf)

Factors affecting the water potential gradient include:
- atmospheric humidity. A high concentration of water vapour in the atmosphere will reduce the water potential gradient between air spaces in the leaf and the atmosphere. The rate of transpiration decreases.
- atmospheric temperature. When temperature increases, the water vapour molecules have more kinetic energy; they move faster away from the stomata as they escape. The rate of transpiration increases.
- Wind moves water vapour away from the stomata as it escapes. This decreases the water potential of the atmosphere and increases the water potential gradient. The rate of transpiration increases.

Factors affecting total stomatal aperture include:
- number of stomata. The more stomata there are per unit area of leaf epidermis, the greater the total aperture and the greater the rate of transpiration.
- light intensity. A rise in light intensity opens stomata, increasing the total aperture and so increasing the rate of transpiration.

examiner tip
Everyone knows that clothes on a washing line dry best when it is warm, dry and windy. The same applies to the loss of water from plants.

Measuring the rate of transpiration

The rate of transpiration can be measured using a **potometer**.

There are two two basic types of potometer.
- The bubble potometer shown in Figure 38 measures the rate of water **uptake** by a leafy shoot by timing how quickly a bubble in a column of water moves a certain distance along capillary tubing (of known diameter) attached to the shoot. The leafy shoot should be placed in the apparatus under water so that no unwanted air bubbles are introduced.
- The mass potometer shown in Figure 39 measures the water loss from a plant by measuring its change in mass over a period of time.

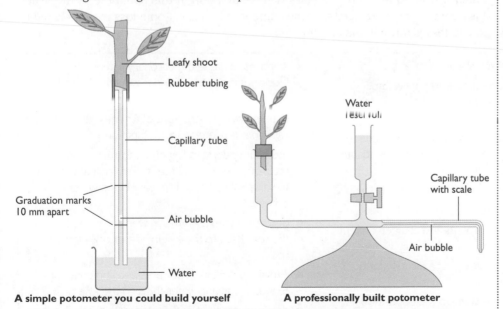

A simple potometer you could build yourself   **A professionally built potometer**

Figure 38 Two types of bubble potometer

The simple potometer in Figure 38 is cheap and easy to assemble, but repeat readings are difficult to obtain as the apparatus must be re-assembled each time. The alternative potometer allows repeat readings to be taken easily. After each reading, more water can be run from the reservoir, pushing the air bubble back to the end of the capillary tube, ready for another reading to be taken.

Both potometers measure water uptake, which we assume is directly related to water loss by transpiration.

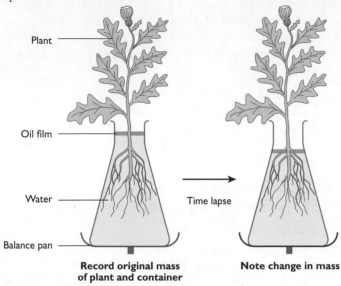

Figure 39 The mass potometer

The apparatus in Figure 39 does actually measure the amount of water lost, rather than the amount taken up. However, its accuracy is limited by the accuracy of the balance used to measure the mass. The assumption in this apparatus is that water loss from the plant accounts for the entire loss in mass. Some loss in mass could be due to losses through the oil film.

Summary

After studying this topic, you should be able to:

- explain the concept of surface area-to-volume ratio and explain its importance when considering biological exchange surfaces
- use your understanding of surface area-to-volume ratio to explain how each of the following is adapted for efficient gas exchange
  - the plasma membrane of a single-celled organism
  - the spiracles and tracheae of an insect
  - the gill lamellae, gill filaments and counter-current ventilation mechanism of a fish
  - the stomata and mesophyll of a dicotyledonous leaf
- interpret information relating to the opposing needs of an efficient gas exchange surface and the limitation of water loss shown by terrestrial insects and xerophytic plants

- identify and name the coronary arteries and the blood vessels entering and leaving the heart, liver and kidneys of a mammal
- explain how the structure of arteries, veins and capillaries is related to their function
- use your understanding of the formation and reabsorption of tissue fluid to explain data relating to values of hydrostatic pressure and water potential
- describe and explain the movement of water from the root hairs to the xylem of a dicotyledonous root
- explain the roles of root pressure and cohesion-tension in moving water through the xylem
- describe, interpret and explain data relating to the effects of light, temperature, air humidity and air movement on transpiration

Different cells in different organisms

Key facts you must know and understand

Different types of cells

You learnt about the different structures of prokaryotic and eukaryotic cells in Unit 1. Plant and animal cells are both eukaryotic, but still show some differences, as shown in the following table.

Structure	Structure present or absent in:	
	Plant cell	Animal cell
Cell wall	Present	Absent
Plasma membrane	Present	Present
Nucleus	Present	Present
Chloroplast	May be present	Absent
Large vacuole	Present	Absent

Even using an optical microscope, we can see the differences between a palisade mesophyll cell from a leaf and an epithelial cell from the small intestine (Figure 40).

Knowledge check 28
What limits the detail of cell structure that we can see using an optical microscope?

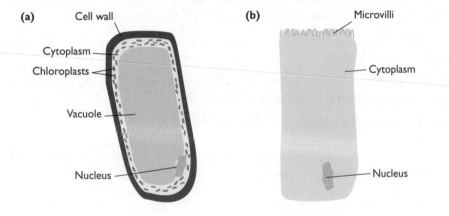

Figure 40 (a) The structure of a palisade mesophyll cell and (b) the structure of an epithelial cell from the small intestine, as seen through an optical microscope

Key structures in plant cells

The cell wall

The cell wall of a plant is a complex structure with several layers. A cell wall always contains the structures shown in Figure 41:
- a middle lamella — made from pectins
- a primary wall — made from cellulose, hemicelluloses and pectin

It may also contain:

- a secondary wall — this may be made from different substances; in xylem cells it is made from lignin

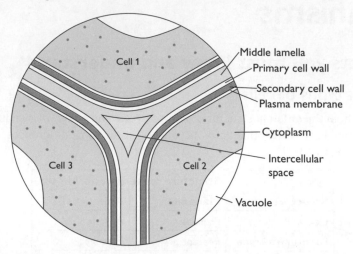

Figure 41 Layers in a cell wall

Knowledge check 29

What is the symplast pathway for water movement in plants?

The cellulose molecules are organised into microfibrils (see page 49), and then into fibres. The fibres in the cell wall run in different directions, conferring greater strength than if they all ran in the same direction, while still allowing some elasticity.

Gaps between the cellulose fibres ensure that the cell wall is freely permeable to molecules of all sizes and form the apoplast pathway for water.

The other molecules in the primary wall, pectins and hemicelluloses, help to hold the cellulose fibres in place.

The lignin in the secondary cell wall of xylem vessels makes the cell wall much more rigid and harder than the cell walls of other plant cells.

The extra rigidity of xylem vessels is important in resisting the inward force of the tension produced by transpiration and is one adaptation to its function of transporting water through a plant. Other adaptations are that:

- xylem vessels are hollow; they 'die' soon after they are formed; all the cytoplasm and organelles are lost
- the end walls of the cell break down

examiner tip

Think how a straw collapses inwards if you suck too hard. Xylem vessels do not do that.

These adaptations allow xylem vessels to form continuous hollow tubes that run throughout the plant, transporting water from root to leaf.

The cell walls of guard cells also adapt them to their function. Figure 42 shows that their walls are thicker in some places than others. When guard cells take in water by osmosis, they swell. The inner, thicker region of the wall cannot stretch; the outer, thinner region stretches so the guard cells curve and open the stoma.

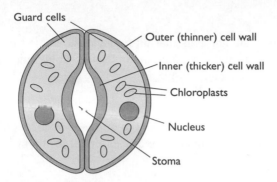

Figure 42 Diagram of guard cells

Chloroplasts

Cells that contain chloroplasts can photosynthesise. The ultrastructure of a chloroplast is linked to harnessing light energy to drive chemical reactions.

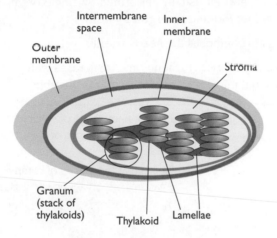

Figure 43 Diagram of a chloroplast

The thylakoid membranes shown in Figure 43 contain chlorophyll and other pigments arranged in such a way that:

- the pigments absorb light energy efficiently
- the energy is used to synthesise ATP and to split water into hydrogen ions, electrons and oxygen
- the hydrogen ions join to an acceptor molecule

Some chemical reactions of photosynthesis take place in the liquid stroma; here:

- carbon dioxide enters a complex cycle of reactions that synthesise glucose
- the hydrogen acceptor molecule and ATP 'drive' this cycle of reactions

Tissues, organs and organ systems

The many different types of cell in animals and plants arise from **cellular differentiation**. Following fertilisation, a zygote divides repeatedly by mitosis. Its new cells become adapted for specific functions — the process of differentiation.

> **examiner tip**
> You will learn about the process of photosynthesis in Unit 4. You will not be tested on it in BIOL2.

Tissues

A tissue is a group of similar cells that all perform the same function. Epithelium, muscle and blood are examples of mammalian tissues. Epidermis, spongy mesophyll and xylem are examples of plant tissues.

Organs and organ systems

Organs are structures within an organism that are made of several different tissues. Each tissue performs its own function and is essential to the overall functioning of the organ.

A muscle (such as the biceps) is an animal organ. It contains **skeletal muscle tissue**, together with arteries and veins (each made from epithelia, smooth muscle and connective tissue), blood and nervous tissue.

A leaf is a plant organ, containing tissues such as epidermis, spongy mesophyll, palisade mesophyll, xylem and phloem.

Major body processes are not usually performed by single organs but by groups of organs working together forming an **organ system**.

The circulatory system comprises the heart, arteries, veins and capillaries.

The breathing system comprises the lungs and trachea, as well as the diaphragm and intercostal muscles that make breathing movements possible.

Knowledge check 30

Explain why an artery is an organ but a capillary is a tissue.

Summary

After studying this topic, you should be able to:

- describe the fundamental differences between plant cells and animal cells
- recognise and name the structures within a leaf palisade cell seen with an optical microscope
- recognise the ultrastructure, and explain the function of, plant cell walls and chloroplasts
- use your knowledge of eukaryotic features in explaining adaptations of plant cells other than a leaf palisade cell
- explain that the bodies of complex, multicellular organisms are organised into tissues, organs and systems

Different molecules in different organisms

Figure 44 shows how differences in DNA lead to different molecules being synthesised.

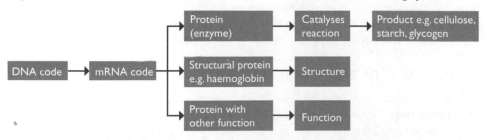

Figure 44 DNA differences produce different molecules

Key concepts you must understand

Haemoglobin

Figure 45 shows that haemoglobin is a protein with a quaternary structure. It is composed of four polypeptide chains, linked together to form a single molecule.

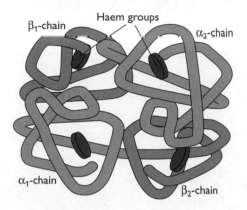

Figure 45 The quaternary structure of haemoglobin

Knowledge check 31

What are the primary, secondary, tertiary and quaternary structures of a protein?

Haemoglobin can bind loosely with oxygen to form **oxyhaemoglobin**; this dissociates easily again to release the oxygen.

The percentage of haemoglobin that has oxygen bound is referred to as the **percentage saturation** of haemoglobin. Figure 46 shows how the percentage saturation varies with the partial pressure of oxygen surrounding the red cells containing the haemoglobin, and is referred to as the **oxygen dissociation curve of haemoglobin**.

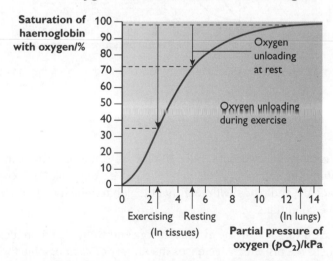

Figure 46 Oxygen dissociation curve of human haemoglobin

Knowledge check 32

If you drew an arrow across Figure 46 to represent the direction of blood flow, would you draw it from left to right or from right to left?

The difference in saturation between the lungs (98%) and the tissues varies depending on activity, and is due to oxygen dissociating from the haemoglobin and being released to the tissues.

The precise structure of haemoglobin depends on the DNA that codes for its amino acid sequence. This differs slightly between organisms and so the different haemoglobins have different structures.

Different haemoglobins have different affinities for oxygen because of their structures. The affinity can also be altered by environmental factors, such as pH and the concentration of carbon dioxide.

The difference in storage carbohydrates in plants and animals is related to the difference in metabolic rates. Animals with a higher metabolic rate than plants have a storage carbohydrate that hydrolyses faster than that in plants.

Key facts you must know and understand

The haemoglobin of animals that live in conditions of low partial pressures of oxygen has a higher affinity for oxygen than normal human haemoglobin, for example:

- on high mountains (the llama)
- in burrows that fill with water (*Arenicola* — a worm that lives in the intertidal zone)
- in the womb (fetal haemoglobin has a much higher affinity for oxygen than adult haemoglobin)

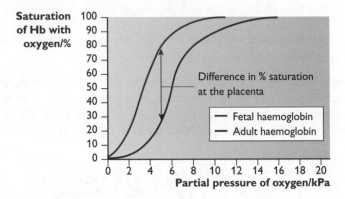

Figure 47 The oxygen dissociation curves of fetal and adult haemoglobin

The dissociation curve of fetal haemoglobin is to the left of the adult haemoglobin curve. This means that it is more highly saturated with oxygen at all partial pressures of oxygen.

Look at Figure 47. At the placenta, only 30% of maternal haemoglobin remains as oxyhaemoglobin. The remaining 70% has released its oxygen. The oxygen diffuses across the placenta and binds with fetal haemoglobin, which can be 75% saturated under the same conditions.

Knowledge check 33

What causes the increased concentration of carbon dioxide in active animals?

When animals are very active, the increased concentration of carbon dioxide in their plasma decreases the pH, which decreases the affinity of haemoglobin for oxygen and more oxygen is released from the oxyhaemoglobin as a result. As Figure 48 shows, an increase in the concentration of CO_2 shifts the dissociation curve to the right.

This is called the Bohr effect and the result of this is to release more oxygen to active tissues than to less active tissues.

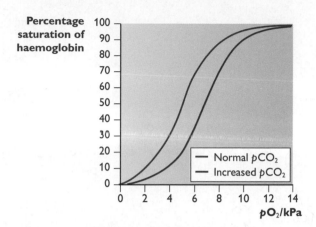

Figure 48 The Bohr effect

Cellulose, starch and glycogen

Cellulose

Cellulose is a structural carbohydrate consisting of many β-glucose molecules joined by condensation reactions to produce an unbranched chain.

Figure 49 shows that cellulose molecules lying side by side can link together by hydrogen bonds to form structures called **micelles**. Micelles are grouped into **microfibrils** and these, in turn, are grouped into larger **fibres** of cellulose.

Knowledge check 34

What is a condensation reaction?

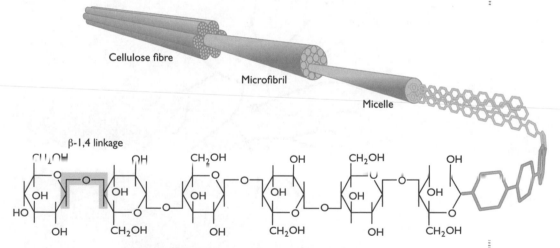

Figure 49 The structure of a cellulose fibre

The fibres of cellulose are used in the synthesis of plant cell walls (see pages 43–45).

Starch

Figure 50 shows that starch contains two polymers of α-glucose:
- amylose — an unbranched chain of α-glucose molecules linked by α-1,4 glycosidic bonds
- amylopectin — a branched chain of α-glucose molecules; in the main chains, molecules are linked by α-1,4 glycosidic bonds, as in amylose, but some α-1,6 glycosidic bonds also occur and these form the branch points

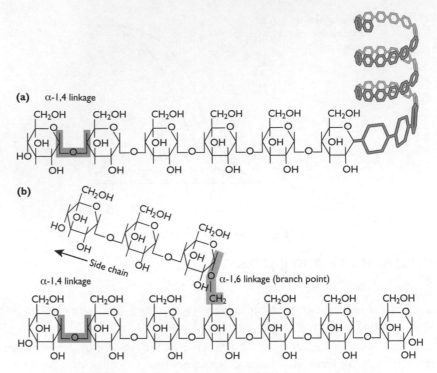

Figure 50 (a) Structure of amylose (b) Structure of amylopectin

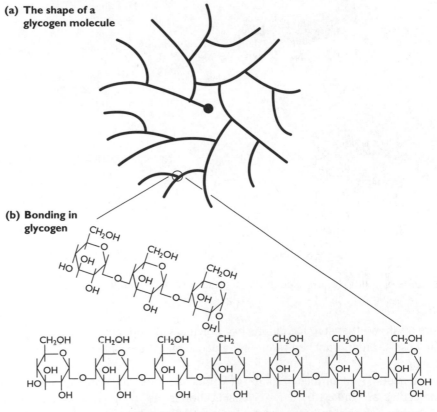

Figure 51 Structure of glycogen

Starch is an effective storage carbohydrate because:

- amylose is a compact molecule, so much can be packed into a starch grain in a cell
- amylopectin is rapidly hydrolysed to glucose, because enzymes can begin to operate on all 'ends' of the branches
- both are insoluble, which means that they will have no osmotic effect on surrounding cells

Glycogen

Animals also store carbohydrate, but as glycogen. Figure 51 shows that glycogen is a very highly branched molecule — much more branched than amylopectin. This allows glycogen to be broken down very quickly, thus helping to support the high metabolic rate of animals.

Knowledge check 35

Why can a highly branched molecule be hydrolysed faster than an unbranched molecule?

After studying this topic, you should be able to:

- explain that haemoglobins are a group of proteins, each with a similar quaternary structure, found in many different animals
- show understanding of the role of haemoglobin in the loading, transport and unloading of oxygen, as shown by its oxygen dissociation curve
- relate the properties of different haemoglobins to the way of life of the animals concerned
- recognise that molecules of cellulose consist of chains of β-glucose linked by glycosidic bonds formed by condensation
- explain how the structures of starch, glycogen and cellulose are related to their respective functions

Summary

Classifying organisms

Key concepts you must understand

Classification systems

Some classification systems of organisms are **artificial classifications**, others are **natural classifications**. They differ in some important ways.

Artificial systems are built on *conveniently observed features*, and groups are *constructed* using the presence or absence of these features. For example, both tarantula and cobra could be placed in the artificial group of 'poisonous animals'. Owls and eagles are both 'birds of prey', but they belong to different natural groups.

Natural systems seek to *discover* groups that exist *as a result of evolution* by looking at as many features of the organisms as possible, with the aim of establishing true *'kinship'* between organisms. For example, the different groups of vertebrate animals seem to be natural groups that are the product of evolution.

Placing organisms in groups based on their evolutionary history is called **phylogeny**. **Phylogenetic trees** show how different organisms have evolved at different times from a common ancestor. Figure 52 shows a phylogenetic tree for some of the main groups of organisms. To compare how closely related groups are, the common branching point must be found. The further back this is, the less closely related the groups are.

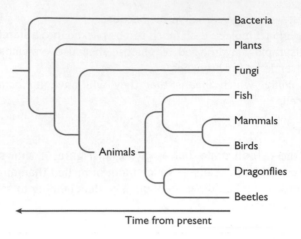

Figure 52 A phylogenetic tree

Each group in a classification system is called a taxon and the study of classification is called **taxonomy**. Modern classification systems are hierarchical, i.e. related taxa are grouped into a larger taxon with no overlap between taxa.

Key facts you must know and understand

The hierarchy of taxa in one current classification system is:

- **Kingdom**
- **Phylum**
- **Class**
- **Order**
- **Family**
- **Genus**
- **Species**

All organisms have a double-barrelled scientific name. It is called a **binomial**. All cats have the same 'first name' or **generic** name (*Felis*) — this is the genus to which they belong. The last name (**specific** name) is different for each species. *Felis catus* is the domestic cat, *Felis silvestris*, the European wild cat and *Felis lynx*, the lynx.

Besides the cats mentioned above, there are other 'cat-like' mammals such as lions, tigers, cheetahs and leopards. The scientific name of the lion is *Panthera leo*. It is in the genus *Panthera* (the panthers). The cheetah, *Acinonyx jubatus*, is in a different genus. However, because these big cats and the smaller cats are really quite similar, they are all placed into one larger group, the **family** of cats — the Felidae.

Cats are clearly different from dogs, yet there *are* similarities. They are both carnivorous mammals and so are grouped into the order Carnivora within the class Mammalia. Mammals belong to the phylum Chordata, which includes all the vertebrates. Finally, chordates belong to the kingdom Animalia.

There are five kingdoms in this classification system:

Kingdom	Organisms included	Main features of organisms
Animalia	All animals	Multicellular, develop from an embryo that forms a hollow ball of cells, have eukaryotic cells, usually motile, usually ingest food into a digestive system
Plantae	All plants	Multicellular, eukaryotic cells with cellulose cell walls, photosynthetic
Fungi	All fungi	Eukaryotic cells, some unicells but most multicellular, non-photosynthetic, cell walls made of chitin, secrete enzymes that digest food outside body
Prokaryotae	Bacteria and blue-green bacteria	Prokaryotic cells with no true nuclei, no membrane-bound organelles, circular DNA, peptidoglycan cell walls
Protoctista	Everything else	Eukaryotic cells, may be motile with no cell walls; may be photosynthetic with non-cellulose cell walls; may be unicellular or multicellular

examiner tip

Be precise when describing the diagnostic features of different kingdoms. For example, cells walls are present in four of the five kingdoms but only plants have *cellulose* cell walls.

Key concepts you must understand

The species concept

The smallest taxon is the species. A species can be defined as a group of organisms that have observable or measurable similarities and are able to produce fertile offspring.

However, this definition has limitations.
- What about species that reproduce asexually? Interbreeding to form fertile offspring cannot occur.
- Some organisms, apparently within one 'species', may be very similar to others, but have different reproductive patterns and so, again, cannot produce fertile offspring.
- Different populations within a species may show some anatomical and physiological differences and may inhabit different geographical or ecological areas. For example, there are two distinct 'sub-species' of the African cheetah.
- The concept cannot easily be applied to fossils.

Some biologists suggest including other criteria in the species concept. These include:
- a common mate recognition system (courtship ritual). This would exclude from a species individuals that appear similar to other individuals, but are **reproductively isolated** from them.
- exploiting the same ecological niche. This is based on the idea that the demands of one particular niche would select just one species to be successful.

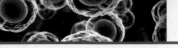

Key concepts you must understand

New techniques used to classify organisms

DNA hybridisation

This technique measures the extent of similarity between two DNA samples. DNA samples that are very similar share many genes and, therefore, many features. They are likely to come from closely related organisms.

The technique measures the extent to which strands of the DNA molecules from the two samples can bind with each other (hybridise).

Protein analysis

Proteins are coded for by DNA. Therefore, similar proteins in two species imply similar DNA. Genetic similarity suggests a close relationship between groups.

Protein analysis aims to compare amino acid sequences of the same protein in different organisms. The more differences, the less closely related the species are presumed to be. However, although data exist for comparison of haemoglobin sequences from many species, the technique is very time-consuming and another technique based on immune responses has been developed.

If a protein from one organism enters another, antibodies specific to that protein will be produced against it. A protein from a third organism, exposed to the antibodies, will only be fully agglutinated or precipitated by the antibodies if the two proteins are very similar. If the two proteins are dissimilar, the extent of the agglutination will be very different also.

Key facts you must know and understand

DNA hybridisation

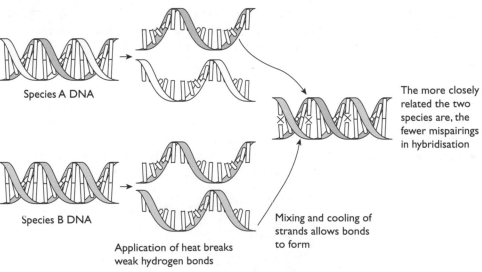

Species A DNA

Species B DNA

Application of heat breaks weak hydrogen bonds

Mixing and cooling of strands allows bonds to form

The more closely related the two species are, the fewer mispairings in hybridisation

Figure 53 Technique of DNA hybridisation

This technique is carried out as follows.

- DNA samples from two different species are heated separately to nearly boiling to separate the strands.
- The two samples are mixed and allowed to cool. This allows the separate strands from the two species to hybridise (re-bind).
- The hybridised strands are reheated until they separate again. The strands with few hydrogen bonds separate at a lower temperature than those with more hydrogen bonds.

Strands from the same species will show 100% hybridisation as they will be completely complementary. Samples from closely related species will show high-percentage hybridisation, whereas it will be lower for more distantly related species.

Using DNA hybridisation techniques, the phylogenetic tree of humans and the great apes appears as in Figure 54.

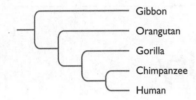

Figure 54 The phylogenetic tree of humans and the great apes, based on DNA hybridisation data

Protein analysis

The immune response comparison technique is carried out as follows.

- A protein from species A is injected into an experimental animal. The animal makes antibodies against it.
- Shortly afterwards, the same protein from species B is exposed to the antibodies and the strength of the new agglutination reaction is estimated.
- If the proteins are very similar, the second agglutination reaction will be stronger than if they are only slightly similar.

Using this immunological technique, the phylogenetic tree of humans and the great apes shown in Figure 55 appears slightly different from that obtained with the DNA hybridisation technique.

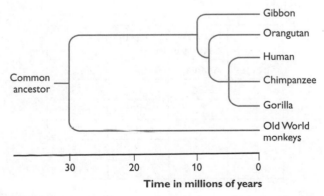

Figure 55 Phylogenetic tree of humans and great apes based on immunological data

Summary

After studying this topic, you should be able to:

- explain the principles and importance of taxonomy
- explain that classification systems consist of a hierarchy in which groups are contained within larger composite groups between which there is no overlap
- interpret phylogenetic trees in terms of evolutionary history
- recognise that one hierarchy consists of the following taxa: kingdom, phylum, class, order, family, genus and species
- use your understanding of the above hierarchy to complete the classification of a given organism
- define the term species and show understanding of the tentative nature of classifying organisms as distinct species
- understand that organisms can be classified on the basis of observable features, by differences in the species-recognition behaviour as shown by their courtship displays, by comparisons of their DNA base sequence and by comparison of the amino acid sequence of their proteins
- describe how the melting points of hybrid DNA and the immunological comparisons of different proteins contribute to classification systems, and interpret data relating to these techniques

Biodiversity

Key concepts you must understand

Measures of biodiversity

The most usual way to think of biodiversity is in terms of **species richness**. This is quite simply the number of different species that are present in an **ecosystem**.

However, if only one or two individuals of a particular species are present in an ecosystem, they won't be contributing a great deal to the biodiversity of the system. A more useful concept is **species diversity**. This takes into account not just how many different species are present, but also the success of each species in the ecosystem.

An **index of diversity** can be calculated and this can be used to give a picture of the ecosystem as a whole.

The examples below relate to three areas containing the same six species (they have the same species richness) and the same total number of organisms — yet the areas are clearly very different.

examiner tip

There are several different indices of diversity. Consequently, it is highly likely that examiners will give you the formula to use when calculating an index of diversity in a BIOL2 paper.

Species	Number of organisms of each species		
	Area 1	Area 2	Area 3
A	86	16	23
B	5	17	25
C	2	16	27
D	3	17	5
E	1	17	12
F	3	17	8

The species diversity of the three areas should reflect the difference in abundance of the six species within each area. An **index of diversity** can be calculated from the formula:

$$d = \frac{N(N-1)}{\Sigma n(n-1)}$$

In this formula, d is the index of diversity, N is the total number of organisms of all species and n is the total number of organisms of each species.

For area 1,

$$d = \frac{100 \times 99}{(86 \times 85) + (5 \times 4) + (2 \times 1) + (3 \times 2) + (1 \times 0) + (3 \times 2)} = 1.348$$

For area 2,

$$d = \frac{100 \times 99}{(16 \times 15) + (17 \times 16) + (16 \times 15) + (17 \times 16) + (17 \times 16) + (17 \times 16)} = 6.314$$

For area 3,

$$d = \frac{100 \times 99}{(23 \times 22) + (25 \times 24) + (27 \times 26) + (5 \times 4) + (12 \times 11) + (8 \times 7)} = 4.911$$

A low value for the index of diversity suggests:
- only a few successful species, perhaps only one
- the environment is quite hostile with relatively few ecological niches and only a few organisms are really well adapted to that environment
- food webs that are relatively simple
- change in the environment would probably have quite serious effects

A higher diversity index suggests:
- a number of successful species and a more stable ecosystem
- more ecological niches are available and the environment is likely to be less hostile
- complex food webs
- environmental change is likely to be less damaging to the ecosystem as a whole; tropical rainforests provide an example of a stable ecosystem with high species diversity

Biodiversity operates on other levels also. Besides species diversity, we should consider:
- genetic diversity — the size of the gene pool of a species (the variety of genes and alleles of a species). If local populations are lost, the genetic diversity will be reduced, even though the overall species diversity remains unaltered.
- ecosystem diversity — an organism may exist in a number of different ecosystems. If one of these is lost, then although that species is not lost (although its gene pool will be reduced), others may be. Certainly, the community of organisms in that ecosystem will be lost. More different ecosystems mean more different habitats and this gives more chance for one species to evolve different varieties.

The impact of human activity on biodiversity

A report by the World Conservation Union concluded that: 'The world's biological diversity is more threatened now than at any time since the extinction of the dinosaurs 65 million years ago'.

Key facts you must know and understand

Deforestation

Deforestation is usually carried out for two main reasons:

- to clear land for human activity, such as mining, agriculture or house building
- to obtain timber to make paper, charcoal, furniture, or as a building material

Tropical rainforest is one of the most complex and species-rich ecosystems in the world. There are several 'layers' to tropical rainforest, as shown in Figure 56.

Figure 56 Structure of a tropical rain forest

Knowledge check 39

Use Figure 56 to define the terms 'canopy' and 'emergents' in relation to a tropical rainforest.

Rainforest covers about 7% of the earth's surface and contains 25% of the known species, most of which are found in the canopy and emergents.

Felling tropical rainforest has far-reaching effects.

- There is a serious reduction in species diversity. Many ecological niches are destroyed when trees are felled and the species that fill these niches are lost.
- There is a reduction in the rate at which carbon dioxide is removed from the atmosphere. In addition, if the trees are burned, then carbon dioxide is added to the atmosphere.
- There is a reduction in the amount of nitrogen returned to the soil as much of the timber is taken from the area. Any tree trunks not removed from the area are slow to decay and the soil is depleted in nitrate for many years.
- If the felled area is allowed to regenerate, shrubs are often the first plants to grow and these out-compete the slower-growing trees for mineral ions and light. The area may not return to a rainforest ecosystem, but a much less complex one.

Knowledge check 40

Nitrate-deficient soil affects plant growth. Name **two** important chemicals found in plants that contain nitrogen.

The felling of trees need not be totally destructive and the practice need not be halted. However, the rainforests must be conserved, and felling and re-planting in a planned cycle over a number of years can do this. This could give a sustainable yield of timber, without endangering the species diversity of the rainforests.

Agricultural practices

Large areas of land given over to the production of just one crop plant (such as maize or oilseed rape) inevitably bring a reduction in biodiversity for several reasons, including:

- the area is dominated by just one species, drastically reducing the number of niches for other organisms
- organisms that might live there are regarded as pests as they reduce the crop yield and are controlled by pesticides
- hedgerows are removed to create bigger fields that accommodate large agricultural machinery; this reduces still further the number of habitats and niches, and therefore the biodiversity of the area
- wetlands are drained to create more agricultural land

Other agricultural practices reducing biodiversity include:

- the widespread use of fertilisers to maintain soil fertility; these can run off into nearby waterways causing eutrophication, a process that leads to the water becoming anoxic (lacking oxygen)
- culling or hunting some species because of their impact on livestock (e.g. badgers and foxes)
- growing more than one crop per year in the same field means that the field almost never has 'stubble' growing, which can be a valuable habitat
- stopping the practice of crop rotation

Traditionally, crops were rotated so that in a particular field a cereal would be grown one year, then perhaps a root crop such as carrots, then a legume such as beans and then perhaps one year 'fallow' (just grass, no crop). The rotation would be carried out with different timings in different fields, so that all crops were always available. This meant that different animals could find different habitats.

Knowledge check 41

Suggest how crop rotation benefits the soil.

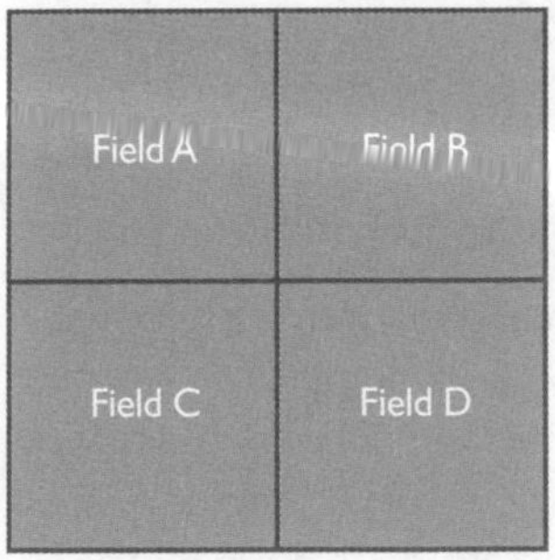

Field	Crop			
	Year 1	Year 2	Year 3	Year 4
A	Root	Legume	Fallow	Cereal
B	Cereal	Root	Legume	Fallow
C	Fallow	Cereal	Root	Legume
D	Legume	Fallow	Cereal	Root

Figure 57 Crop rotation

After studying this topic, you should be able to:

- calculate an index of diversity using the formula $d = N(N-1)/\sum n(n-1)$
- show understanding of the way in which deforestation and agriculture affect biodiversity
- interpret data relating to the effects of human activity on species diversity
- evaluate the benefits and risks associated with the effects of human activity on species diversity

Summary

Questions & Answers

This section contains questions similar in style to those you can expect to see in BIOL2. The limited number of questions in this guide means that it is impossible to cover all the topics and all the question styles, but they should give you a flavour of what to expect. The responses that are shown are real students' answers to the questions.

There are several ways of using this section. You could:

- 'hide' the answers to each question and try the question yourself. It needn't be a memory test — use your notes to see if you can actually make all the points you ought to make
- check your answers against the students' responses and make an estimate of the likely standard of your response to each question
- check your answers against the examiner's comments to see where you might have failed to gain marks
- check your answers against the terms used in the question — for example, did you *explain* when you were asked to, or did you merely *describe*?

Examiner's comments

Each question is followed by a brief analysis of what to watch out for when answering the question (shown by the icon ⓔ). All student responses are then followed by examiner's comments. These are preceded by the icon ⓔ and indicate where credit is due. In the weaker answers, they also point out areas for improvement, specific problems, and common errors such as lack of clarity, weak or non-existent development, irrelevance, misinterpretation of the question and mistaken meanings of terms.

Tips for answering questions

Use the mark allocation. Generally, one mark is allocated for one fact, concept or item in an explanation. Make sure your answer reflects the number of marks available.

Respond appropriately to the command words in each question, i.e. the verb the examiner uses. The terms most commonly used are explained below.

- **Describe** — this means 'tell me about…' or, sometimes, 'turn the pattern shown in the diagram/graph/table into words'; you should not give an explanation.
- **Explain** — give biological reasons for *why* or *how* something is happening.
- **Calculate** — add, subtract, multiply, divide (do some kind of sum!) and show how you got your answer — *always* show your working!
- **Compare** — give similarities *and* differences between…
- **Complete** — add to a diagram, graph, flowchart or table.
- **Name** — give the name of a structure/molecule/organism etc.
- **Suggest** — give a plausible biological explanation for something; this term is often used when testing understanding of concepts in an unfamiliar context.
- **Use** — you must find and include in your answer relevant information from the passage/diagram/graph/table or other form of data.

Question 1 **The structure and function of DNA**

(a) Figure 1 represents the structure of the DNA molecule.

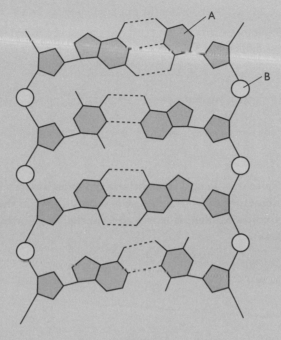

Figure 1

(i)   Name the structures labelled **A** and **B**.   (2 marks)

(ii) Use the diagram to explain why the DNA molecule is sometimes described as consisting of two polynucleotide strands.   (1 mark)

(b) 15% of the bases in a sample of DNA are adenine. What percentage will be guanine? Explain your answer.   (2 marks)

Total: 5 marks

ⓔ This is the sort of question that might appear early in BIOL2, targeted at an E-grade candidate. When naming a structure, as in question (a)(i), give a precise name. If you cannot remember it, move on to the next part of the question. In (a)(ii), you *must* refer to the diagram to gain full marks. Part (b) tests your understanding of base pairs; the calculation itself is easy.

Student A

(a) (i)   A — nucleotide **a**; B — phosphorus **b**

    (ii) There are two strands with lots of nucleotides joined. **c**

(b) 15%. They are complementary bases. **d**

ⓔ **1/5 marks awarded a** The student has confused the terms nucleotide and nitrogenous base and so fails to score the first mark. **b** Phosphorus is a chemical element and so is not an acceptable answer as B is a phosphate group. **c** Correct. **d** The reasoning is incorrect and inevitably so is the answer.

Student B

(a) (i) A is a nitrogenous base **a** (adenine, thymine, cytosine or guanine); B is a phosphate **a** group which links adjacent nucleotides.

(ii) Each strand consists of many nucleotides linked by the phosphate groups **b**. 'Poly-' means many, like in a polygon.

(b) 15% adenine is complementary to thymine, so there will be 15% thymine also. This adds up to 30%. So 70% will be cytosine and guanine together **c**. They are complementary bases, so there will be equal amounts of each. 70/2 = 35 **d**.

ⓔ 5/5 marks awarded **a** The student has provided more information than just the name but gains both marks. **b** This sentence alone gains the available mark. **c** This logic is correct. **d** The student has not added the percentage symbol to the answer but, since the question asked for a percentage, an answer of 35 gains the marks.

ⓔ **Despite this being a straightforward question, testing basic understanding of the structure of DNA, student A scores only 1 of the 5 marks, a grade-U performance. Student B scores all 5 marks, as you would expect of an A/B-grade candidate.**

Question 2 **Biodiversity**

A scientist recorded the numbers of several types of plant found in each of two areas. Her results are shown in the table.

Plant species	Number of plants (n) in area A	Number of plants (n) in area B
Woodrush	4	2
Holly	8	3
Bramble	3	3
Yorkshire fog	6	12
Sedge	7	4
Buttercup	7	4
Total (N)	35	28

(a) Use the formula

$$d = \frac{N(N-1)}{\Sigma n(n-1)}$$

to calculate the diversity index for each area. (3 marks)

(b) Give three deductions that might be made about an area with a low diversity index. (3 marks)

(c) Explain two ways in which agricultural practices can reduce biodiversity. (4 marks)

Total: 10 marks

ⓔ You are given the formula to use in the calculation and told in the table what the symbols n and N represent, so (a) is a matter of using the numbers correctly. Don't forget you can gain marks if your working is correct even if your final answer is wrong. Questions (b) and (c) test AO1 and are not related to the table. If you do not like calculations, you could answer (b) and (c) first and come back to (a) at the end of the examination when you are feeling more confident.

Student A

(a) For area A,

$$d = \frac{N(N-1)}{\Sigma n(n-1)} = \frac{35 \times 34}{(4 \times 3) + (8 \times 7) + (3 \times 2) + (6 \times 5) + (7 \times 6) + (7 \times 6)} = 6.33$$

Area B

$$= \frac{28 \times 27}{(2 \times 1) + (3 \times 2) + (3 \times 2) + (12 \times 11) + (4 \times 4) \, \mathbf{a} + (4 \times 4) \, \mathbf{a}} = 4.25$$

(b) A low index of diversity means the area does not have many species present **b**. Therefore it will only have simple food chains and if something alters the environment, the whole ecosystem might collapse.

(c) Creating large fields growing just one crop reduces biodiversity because there are fewer habitats **c** and crop rotation does also **d**.

ⓔ **6/10 marks awarded a** There is 1 mark here for each correct answer and 1 mark for evidence of a correct method. This student clearly understands how to carry out the procedure,

but in the second calculation has made a slip. The last two items of the divisor should be (4×3) and not (4×4) as written. This gives a wrong answer and the student obviously will not gain that mark. By showing the correct method, however, this student is credited with the 'working mark' and scores 2 of the 3 available marks. **b** This statement is not quite accurate. There could be a good number of species present. It is the idea of the area being dominated by only a few that is not made clear. The student scores 2 marks for statements about the simplicity and stability of the ecosystem. **c** The student scores 2 marks for correctly identifying one practice and explaining its effect. **d** Crop rotation does not reduce biodiversity; stopping crop rotation reduces biodiversity. Perhaps this is what the student meant, but an examiner can only mark what is written.

Student B

(a) The index of diversity for Area A is 6.33. The index for Area B is 4.45. **a**

(b) A low index of diversity indicates an area dominated by just a few species **b**. The area could be quite hostile **b** and the ecosystem unstable **b**.

(c) Removing hedgerows reduces the number of habitats. **c** Also, clearing wetlands reduces biodiversity. **d**

e **9/10 marks awarded** **a** As the student has both answers correct, s/he must have used the formula correctly and gains all 3 marks. Had this student made the same slip as student A, s/he would probably have scored only 1 mark, as no working is shown. **b** These three points score 3 marks. **c** The student gains 2 marks for identifying and explaining one practice. **d** 1 mark for identifying a second practice but there is no explanation.

e **Student A scores 6 marks overall (grade C), but with a little more care and attention to detail could probably have scored more. The slip in the calculation in (a) and careless wording in (c) cost at least 2 marks. Student B scores 9 of the 10 marks — a grade-A performance.**

Question 3 **Mitosis and meiosis**

Figure I shows a cell in a stage of mitosis. The cell contains just two pairs of homologous chromosomes.

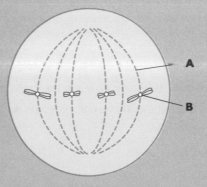

Figure 1

(a) (i)   **What are *homologous chromosomes*?**   (1 mark)

   (ii) **Identify the structures labelled A and B on Figure I.**   (2 marks)

   (iii) **Name the stage of mitosis represented in this diagram. Give a reason for your answer.**   (1 mark)

Figure 2 shows the life cycle of a mammal.

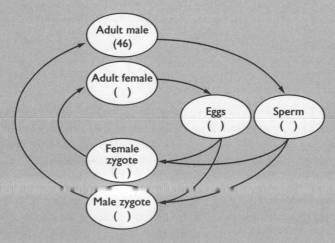

Figure 2

(b) (i)   **Mark on the diagram one stage where meiosis takes place and one place where mitosis takes place.**
                                                     (2 marks)

   (ii) **Complete the empty boxes to show the number of chromosomes per cell.**   (1 mark)

Total: 7 marks

🔴 Parts (a)(i) and (ii) clearly test AO1 and the mark tariff is 1 mark per correct response. To gain the mark in (a)(iii), you must tell the examiner *how* you identified the stage of mitosis; simply naming it will not be enough — you might have guessed the name without understanding mitosis at all. Part (b) involves interpretation of a diagram and tests your understanding of when mitosis and meiosis occur. Candidates often forget to answer questions that involve writing something on a diagram instead of along a dotted line, so watch out for this.

Student A

(a) (i) They have the same genes in the same order. **a**

  (ii) **A** is the spindle; **B** is a chromosome. **b**

  (iii) Prophase **c**

(b) (i) and (ii)

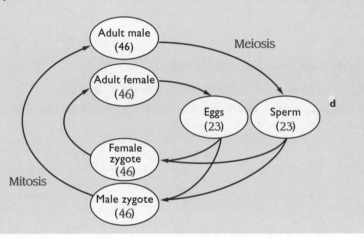

🔴 **5/7 marks awarded a** Correct. **b** This student has not looked carefully enough at label **B**, which points to the centromere and so scores only 1 mark. **c** This is incorrect and the student has also failed to give the required explanation. **d** The labels for part (i) are clear and the student scores these 2 marks; (b)(ii) is also correct and scores the mark.

Student B

(a) (i) A homologous pair of chromosomes is **a** a pair of chromosomes that have the same genes along their length **b** (although they may not have the same alleles).

  (ii) **A** is the spindle or, more accurately, one of the spindle fibres. **c**

  B is the centromere **d**, which holds the chromatids in a chromosome together.

  (iii) Metaphase **e**

(b) (i) and (ii)

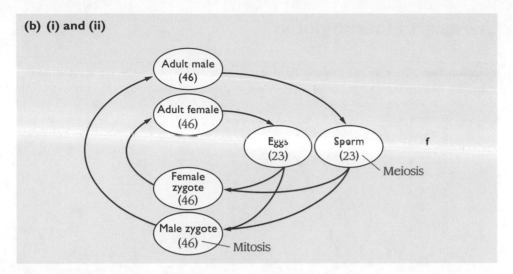

🅔 **5/7 marks awarded a** The student has written out part of the question again. This does not gain marks and wastes time. **b** This answer is correct and gains the mark. **c** 'Spindle' or 'spindle fibre' would gain the mark; again the student has written more than was needed. **d** The correct answer. **e** The correct answer but no explanation, so no mark. **f** The labelling is unfortunate in (b)(I). Meiosis does not actually occur in the sperm, so this mark is not awarded. However, the zygote does divide by mitosis and so this mark can be awarded. The answer to (b)(II) is correct and scores the mark.

🅔 **Student A scores 5 marks overall, as does student B. This is a fairly undemanding question, involving recall of some basic knowledge of mitosis and a fairly simple application of the knowledge of mitosis and meiosis to life cycles. Both students show B-grade standard.**

Question 4 **Haemoglobin**

Oxygen is transported in mammalian blood by the protein haemoglobin in red blood cells. Figure 1 shows the oxygen dissociation curve for human haemoglobin at two different values of pH.

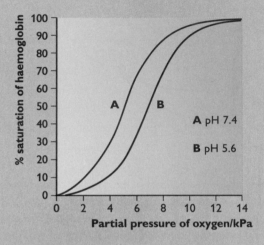

Figure 1

(a) Explain what is meant by 'the percentage saturation of haemoglobin'. (1 mark)

(b) Explain how vigorous exercise produces a dissociation curve similar to curve **B**. (2 marks)

(c) Add to the graph the dissociation curve you would expect for an animal that lives in an area with a low partial pressure of oxygen. Explain the curve you have drawn. (3 marks)

Total: 6 marks

ⓔ This question tests your understanding, rather than just recall, of this topic. Part (a) asks for a definition but part (b) requires you to provide biological reasons for something. Watch out in (c) where marks will be awarded for giving reasons why you drew the curve where you did — simply getting the curve in the right place will only get you 1 of the 3 marks.

Student A

(a) The amount **a** of haemoglobin carrying oxygen.

(b) Vigorous exercise would make the haemoglobin more acidic **b** and so it would not carry oxygen as well. Curve B is lower than curve A. **c**

(c)

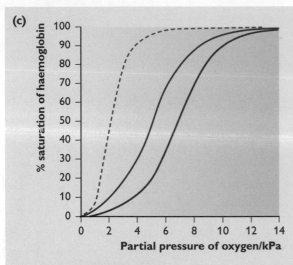

Partial pressure of oxygen/kPa

The haemoglobin of animals like the llama has a lower affinity **d** for oxygen than human haemoglobin, so the curve is shifted to the left **e**.

ℰ 1/6 marks awarded a 'The amount of haemoglobin carrying oxygen' is vague. It could mean 10 g or 78 moles or any other 'amount'. Percentage saturation must convey the idea of proportion; 'the amount of haemoglobin bound to oxygen in 100 cm³ of blood' would be acceptable. **b** The student does not seem to understand that the pH values referred to in the question are plasma pHs. **c** This is a description; the question required an explanation. **d** The student gains 1 mark for the new curve but is clearly confused about high and low affinity and fails to gain the second mark. **e** The comment about shifting the curve to the left does not add anything new to the graph s/he has drawn.

Student B

(a) The percentage saturation of haemoglobin is the proportion of haemoglobin in a certain volume of blood **a** that is actually carrying oxygen, that is, oxyhaemoglobin.

(b) In vigorous exercise, a lot more carbon dioxide is released **b** from the extra respiration. Carbon dioxide results in carbonic acid forming in the plasma, **b** which will lower the pH of the plasma. Also, lactate (lactic acid) may be produced which would also lower the pH of the plasma. This change in the curve is called the Bohr effect or the Bohr shift.

(c)

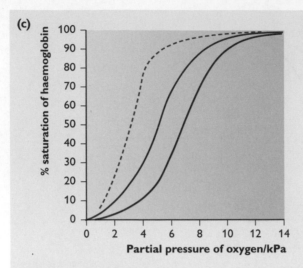

Animals like the llama live on mountains where there is less oxygen. Their haemoglobin has a higher affinity **c** for oxygen than human haemoglobin and so is able to load oxygen even when there is not much around. **c**

ℯ **6/6 marks awarded a** The correct answer. **b** Both these points provide a clear explanation and gain the 2 marks. **c** 1 mark is awarded for the curve, and these two points provide a clear justification and score the remaining 2 marks.

ℯ **Student A scores only 1 mark (grade U) while student B scores all 6 (grade A) for this question. Student A has failed to gain some marks unnecessarily. Although unclear about the concept of 'affinity for oxygen', s/he probably knew that the llama's haemoglobin was adapted to load oxygen under low oxygen tensions and should have said so, rather than just re-stating what had been drawn in the graph. Be careful not to do this.**

Question 5 **Variation**

(a) A student collected and weighed seeds from several pea plants. Their masses are shown in the table.

Mass/g	Number of seeds
< 1.0	0
1.1–1.5	1
1.6–2.0	3
2.1–2.5	7
2.6–3.0	11
3.1–3.5	5
3.6–4.0	2
> 4.0	0

(i) Plot a graph of these results.   (3 marks)

(ii) What sort of variation is shown by these data?   (1 mark)

(b) The student germinated and grew these seeds in a greenhouse and measured the heights of the plants after 42 days. His results are shown in the graph.

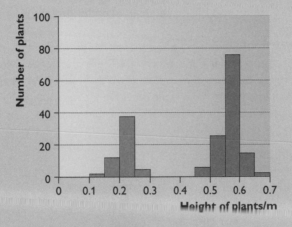

(i) Suggest why he grew the plants in a greenhouse rather than in the open.   (2 marks)

(ii) Describe and explain the variation shown between the pea plants.   (4 marks)

Total: 10 marks

ⓔ Your ability to plot a graph is usually tested in BIOL3, so it is highly unlikely that it will be tested in BIOL2 as well. Marks are awarded for choosing the correct type of graph, using and labelling the axes correctly, choosing an appropriate scale and plotting the data accurately. Part (b)(i) is a test of AO3 skills and will reward you for any reasonable suggestion — for 2 marks you also need to provide an explanation for your suggestion. Notice (b)(ii) asks for both a description (turn the graph into words that describe the trend) and an explanation (reasons for the trend).

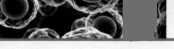

Student A

(a) (i)

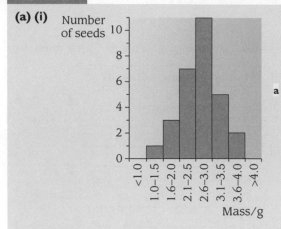

(ii) Categoric **b** variation

(b) (i) If you grow the plants in the open, you don't know what will happen, but in the greenhouse, you can keep it the same temperature **c** all the time.

(ii) There are two definite groups of plants **d**, so these probably have different genes **e**. But all the plants in each group aren't the same. Some could be getting more sunlight than others **f** which would make them photosynthesise better and this would lead to better growth **g**.

ⓔ **8/10 marks awarded a** The student has drawn a histogram with all values plotted correctly and scores 3 marks. **b** This is incorrect. **c** The first sentence is too vague to gain marks but the reference to controlling temperature is worth 1 mark. **d** An adequate description gains 1 mark. **e** This is loose terminology — the genes are the same although the alleles of each gene might not be — but the concept is understood and the mark awarded. **f,g** These explanatory points gain 2 marks.

Student B

(a) (i)

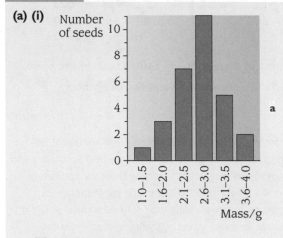

(ii) Continuous variation

(b) (i) In a greenhouse, you can control the conditions **b** and so you know that any differences are due to differences in the plants themselves. **b**

    (ii) There are genetic differences **c**, as there are two distinct groups **c**. But the variation within a group **c** is due to the environment. **c**

ⓔ **8/10 marks awarded a** The student has assigned the x and y axes correctly but, after that, has made several errors and only scores 1 mark. S/he has drawn a bar chart, when a histogram was appropriate (bars touching, as in student A's graph). Also, the graph is incomplete as no values are plotted for <1.0 and >4.0. Both these values are 0, but they should have been plotted. **b** These two points gain both marks. **c** These four points score all 4 marks.

ⓔ **Student B made silly mistakes in plotting the graph and student A should really have known, after correctly plotting a histogram, that the variation shown was continuous. Despite this, both students score 8 marks (grade A).**

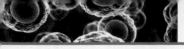

Question 6 **Classification**

Amoeba proteus and *Paramecium caudatum* are both unicells. They both have eukaryotic cells and are motile.

(a) (i) Complete the table showing the classification of *Amoeba*. (2 marks)

Taxon	Taxonomic group
Kingdom	
	Plasmodroma
	Rhizopoda
	Amoebida
	Amoebidae
Genus	
Species	

(ii) Name *two* taxa that the two organisms could share. (1 mark)

(iii) Name *one* taxon that the two organisms do not share. (1 mark)

(b) Give two differences between these cells and bacterial cells. (2 marks)

Total: 6 marks

 Part (a)(i) will be marked as 1 mark per correct column. You need a simple mnemonic to remember the taxonomic groups as KPCOFGS. Clues in the description should help you decide which kingdom these two organisms belong to. The names of the genus and species are given to you in the question — yes, it really is that simple. Parts (a)(ii) and (iii) test your understanding of a taxonomic hierarchy. This involves such a low level of understanding that the mark tariff is low. Part (b) is straight recall (AO1) of two differences between prokaryotic and eukaryotic cells. You might choose to answer (b) first because it does not involve sorting out the information in the rest of the question.

Student A

(a) (i)

Taxon	Taxonomic group
Kingdom	Animalia **b**
Phylum	Plasmodroma
Class	Rhizopoda
Order	Amoebida
Family **a**	Amoebidae
Genus	Amoeba
Species	proteus

(ii) They are both in the Animalia **c** and the Plasmodroma.

(iii) They are in a different species. **d**

(b) They have a nucleus **e** and can move **f**.

@ **4/6 marks awarded a** The first column in the table is correct and scores 1 mark. **b** The student has placed *Amoeba* in the kingdom Animalia, which is wrong. **c** The student has already failed to gain a mark for naming the kingdom incorrectly, so will not be penalised again for the same error. An examiner will accept that the student has told us that the taxa are kingdom and phylum, so the mark is awarded. **d** This is correct and gains the mark. **e** This gains 1 mark but **f** is incorrect — some bacteria have flagella, which they use to move.

Student B

(a) (i)

Taxon	Taxonomic group
Kingdom	Protoctista
Phylum	Plasmodroma
Class	Rhizopoda
Order	Amoebida
Family	Amoebidae
Genus	Amoeba
Species	proteus

    (ii) Kingdom and phylum **a**

    (iii) Species **b**

(b) Eukaryotic cells have true nuclei and membrane-bound organelles **c**, whereas prokaryotic cells don't.

@ **6/6 marks awarded** The student has correctly completed both columns in the table and **a** has answered (a)(ii) correctly. **b** This is correct; note the brevity of the answer, which saves time. **c** Both features are correct, gaining 2 marks.

@ **This is a straightforward question on classification and candidates should score well. Student A scores 4 marks (grade C) and student B scores 6 (grade A).**

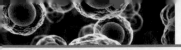

Question 7 Genetic bottlenecks

The graph shows the effect of a genetic bottleneck on the numbers and genetic variability of a population.

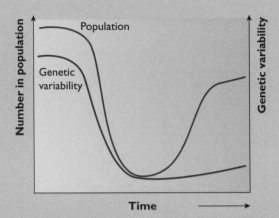

(a) Describe two pieces of evidence in the graph that show that a genetic bottleneck took place. (2 marks)

(b) Explain two possible reasons for the slight increase in genetic variability after the genetic bottleneck. (4 marks)

(c) Explain why the population after the genetic bottleneck is more vulnerable to environmental change than the population before the bottleneck. (4 marks)

Total: 10 marks

ⓔ You are given a conclusion in part (a) and asked to find two pieces of evidence from the graph that support it. Note, this time you are asked only to *describe* the evidence, not to explain how it supports the conclusion. Part (b) offers 2 marks for each of two reasons. Part (c) does not rely on the graph at all — it is a straight test of understanding (AO1).

Student A

(a) The population fell **a** and the genetic variability fell as well **a**.

(b) Mutations **b** would introduce some genetic variability and new combinations of the genes **c** already there would as well.

(c) Because there is less variation, if the environment changes **d**, they will either all be suited to the new environment or they all won't **d**.

ⓔ **5/10 marks awarded a** The student has just done enough to score both marks. **b** This scores 1 mark. **c** The student has confused 'genes' with 'alleles' and so is not awarded the second mark. **d** These two points score 2 marks.

Student B

(a) There was a population crash, but the numbers recovered **a**, whereas the genetic variability fell and stayed low **b**.

(b) Mutations **c** will actually produce new alleles **d**, so increasing the genetic variability. The other way is by random mating and random fertilisation **e**, which will produce new combinations of genes **f** that already exist.

(c) The lack of genetic variation means that if the environment changes **g**, few if any will be adapted to the new condition **h** and most won't survive **i**.

ⓔ **8/10 marks awarded a** and **b** make both points and score 2 marks. **c** This scores 1 mark. **d** This *explains how* mutations increase genetic variability and scores 1 mark. **e** A correct description of how variation can increase for 1 mark, but **f** by referring to new combinations of genes (rather than alleles), the student fails to gain the fourth mark. **g–i** show the 3 marks gained in this answer.

ⓔ **Student A scores 5 marks (grade D), student B scores 8 (grade B). This is a question where understanding of concepts is important, but the detail must also be supplied.**

Question 8 Passing on the genetic material

The graph shows the changes in the amount of DNA in a cell during different stages of the cell cycle.

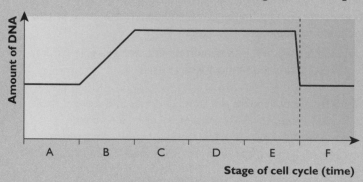

(a) Explain the change in the amount of DNA during:

   (i) stage B   (2 marks)

   (ii) stage E   (2 marks)

(b) (i) Name the stage of the cell cycle labelled C.   (1 mark)

   (ii) Describe the events that take place in the stage labelled A.   (2 marks)

Total: 7 marks

ⓔ In part (a), you will not gain marks for writing that the amount of DNA doubled in stage B and halved in stage E — these are descriptions and the question asks for explanations. Many candidates remind themselves that they should give an explanation by writing the word 'Because' on the first line of the answer space before they begin to answer. After you have worked out from the graph which stages are represented by B and E, you can deduce the other stages in part (b) — that is why the question was written in this sequence.

Student A

(a) (i) The chromosomes had duplicated themselves into chromatids. **a**

   (ii) The cell has divided into two **b**, so some **c** of the DNA ends up in each cell.

(b) (i) Interphase **d**

   (ii) In this stage, the cell is getting ready to divide by growing **e** and making more cell organelles **f** that can be shared between the cells after cell division.

ⓔ **5/7 marks awarded a** The student has the right idea but fails to refer to DNA, which was what the question was about, thus scoring only 1 mark. **b** This is correct and gains 1 mark. **c** 'Some of the DNA' is too vague; it is clear from the data that the amount of DNA is halved. **d** This is correct — the specification refers to interphase in the cell cycle. **e** Two valid points gain 2 marks.

Student B

(a) (i) This is the S phase **a** of the cell cycle. The DNA undergoes semi-conservative replication **a**.

   (ii) This is cytokinesis **b** and when the cell divides, half the DNA **b** ends up in each cell.

(b) (i) G2 **c**

   (ii) In preparation for mitosis, the cell is growing. **d**

ⓔ **6/7 marks awarded a** These two points gain 2 marks. **b** These two points gain 2 marks. **c** This answer is more precise than students A's response but gets the same 1 mark. **d** This answer gains 1 mark.

ⓔ **Overall, student A scores 5 marks (grade C) and student B scores 6 (grade B).**

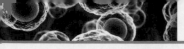

Question 9 Selection in bacteria

The diagram shows two processes by which bacteria transfer DNA.

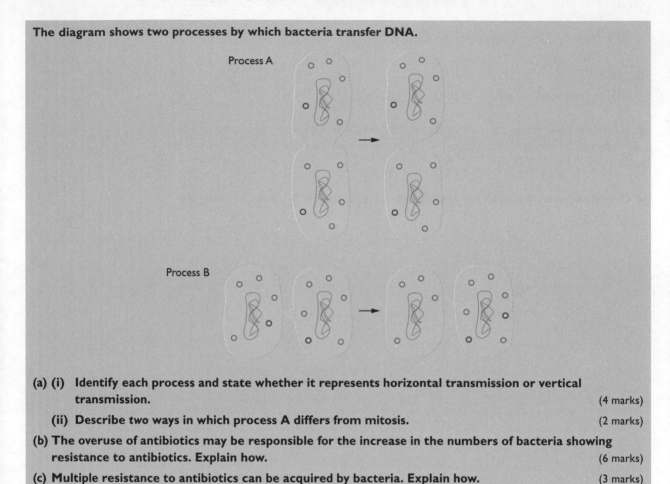

(a) (i)   Identify each process and state whether it represents horizontal transmission or vertical transmission.   (4 marks)

  (ii) Describe two ways in which process A differs from mitosis.   (2 marks)

(b) The overuse of antibiotics may be responsible for the increase in the numbers of bacteria showing resistance to antibiotics. Explain how.   (6 marks)

(c) Multiple resistance to antibiotics can be acquired by bacteria. Explain how.   (3 marks)

Total: 15 marks

ⓔ Part (a)(i) will reward you with 1 mark for each identification and 1 mark for correctly writing vertical or horizontal transmission. 4 marks is a good return for about one minute's work. Part (a)(ii) relies on your understanding why binary fission in bacteria cannot be called mitosis. Part (b) has a tariff of 6 marks — the examiner wants you to write six things you have learnt in your AS course about selection for antibiotic resistance; this will take you about 5–7 minutes. You can gain all 6 marks by using bullet points — you do not have to write in continuous prose if you do not want to. Part (c) is straight recall.

Student A

(a) (i)   A is mitosis **a** because one cell just divides into two. B is fertilisation **a**.
        A is vertical transmission **b** and B is horizontal transmission **b**.

  (ii) This process halves the amount of DNA and introduces variation. **c**

(b) The bacteria adapt to the use of antibiotics by changing to be resistant **d** to them. This makes them survive better **e** so that more of them are resistant.

(c) Bacteria can swap DNA. So one bacterium can give a resistance gene to another bacterium. If it already had resistance to one antibiotic, it is now resistant to two. **f**

ⓔ **4/15 marks awarded a** The student has wrongly identified both A and B but **b** scores 2 marks for correctly identifying vertical and horizontal transmission. **c** When s/he read this question, why did Student A not go back to her/his answer to (a)(i) and change it? The wording of this question makes clear that process A *cannot* be mitosis. This can happen in an examination — one question throws light on your answer to another question — so be on the lookout for it. **d** This answer reads as though the bacteria are 'choosing' to adapt. If this is what the examiner believes you are saying, you will fail to gain marks, as mutations occur by chance regardless of whether antibiotics are present or not. **e** The idea of differential survival is just about there and this gains 1 mark. **f** Although badly expressed, the whole answer shows understanding of the concept and gains 1 mark.

Student B

(a) (i) A — binary fission **a**. This is vertical transmission **b** because it passes from one generation to the next. B is conjugation **a**, and this is horizontal **b** transmission.

(ii) No chromosomes are involved. **c**

(b) Antibiotics cause mutations **d** that make some of the bacteria resistant to the antibiotic anyway. When the antibiotic is used a lot, these bacteria have an advantage **e** because they won't be killed, so they survive **f** and reproduce **g**.

(c) Antibiotic resistance is usually found in genes in plasmids **h**. Bacteria can swap plasmids **i** and transfer resistance to another bacterium.

ⓔ **10/15 marks awarded a** The two processes are correctly identified. **b** These are correct, for another 2 marks. **c** The student identifies one essential difference and scores 1 mark, but surprisingly, having got this point, does not mention that no nucleus is involved or that no spindle is involved. **d** The student is clearly stating that antibiotics cause the mutation, which is completely wrong. **e–g** show correct understanding and gain 3 marks. **h** Identifying the position of the gene gains 1 mark. **i** This gains 1 mark.

ⓔ **Like all questions, this one is only easy if you know the answers, but there is really no excuse for student A's grade-U performance (just 4 marks out of 15). There is nothing devious about any of the questions, and any student who has revised the topic thoroughly should be able to equal student B's score of 10 (grade A/B). If you do not know the material, you cannot hope to score the marks.**

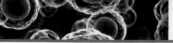

Question 10 Mitosis and tumour formation

Scientists collected data on the most common cancers in men and women in the UK. Their results are shown in the tables.

Site (men)	% of all cancers (men)
Lung	21
Skin	14
Prostate	10
Bladder	5
Colon	6
Stomach	5
Rectum	5
Lymph nodes	3
Oesophagus	2
Pancreas	2
Other cancers	27

Site (women)	% of all cancers (women)
Breast	19
Skin	11
Lung	8
Colon	6
Stomach	3
Ovary	3
Cervix	3
Rectum	3
Uterus	2
Bladder	2
Other cancers	40

(a) (i) The data suggest that there are more cancers of the lung in men than in women. This is not necessarily the case. Explain why.

(2 marks)

(ii) Suggest and explain *two* reasons why the incidence of skin cancer appears to be higher in men than in women.

(4 marks)

(b) The tumour suppressor gene p53 is one of several genes that control mitosis. This gene is mutated in many human cancer cells. Scientists are trying to find ways of repairing the damaged gene. In one trial, they are using altered viruses to carry normal p53 genes into cancer cells. The treatment has shown promising results in laboratory animals.

   (i) Why would introducing normal p53 genes into the cancer cells be an effective way of treating the cancer?

        (3 marks)

   (ii) Give *scientific* reasons why the use of laboratory animals in this research may not be justified.

        (2 marks)

   (iii) The use of viruses to deliver the p53 gene in humans may not be effective. Suggest two reasons why.

        (4 marks)

        Total: 15 marks

ⓔ This is the type of question that appears at the end of BIOL2 — testing your ability to appraise the work of other scientists (AO3). When answering (a)(i), think about the form in which the numbers in the tables are given. In (a)(ii) you will gain 1 mark for each relevant suggestion and a further 1 mark for an appropriate explanation of each suggestion. The mark tariff shows you need to write three different points in your explanation for (b)(i), two reasons in (b)(ii) and an explanation for each of your two reasons in (b)(iii).

Student A

(a) (i) Because we don't know how many cancers there are. **a**

   (ii) Men could spend longer in the sun **b** and could also not look after their skin **c** as well as women.

(b) (i) The normal p53 gene would stop the cancer developing further. **d**

   (ii) The animals may suffer **e** because they are made to develop cancers. Also, because they are different from humans **f**, the results may not help us much.

   (iii) The viruses may not be able to get into the cells **g** and the immune system may destroy them **g**.

ⓔ **4/15 marks awarded a** The student does not make clear that s/he is talking about all cancers. **b** The link with exposure to the sun has been made and scores 1 mark but **c** the suggestion of 'not looking after their skin' is far too vague to be credited. **d** This states little more than was given in the question. **e** The student has started with an emotional response, when the question asked for scientific reasons, but **f** gains 1 mark for this poorly worded reference to the different reactions of different species. **g** 1 mark is gained for each of these suggestions.

Student B

(a) (i) It does not tell us how many cancers there were in men and women, only their percentages **a**. If there were a lot more cancers in women then 8% could give a bigger number **b** than 21% in men.

(ii) Prolonged sunbathing could result in sunburn **c**, which could lead to cancers forming. Also they may be more at risk because of where **d** they work.

(b) (i) The p53 gene is a tumour suppressor gene. Genes like this stop or slow down the development **e** of a tumour. So putting the normal gene into cancer cells would stop them dividing **e**.

(ii) Mice are a different species **f** and so may respond differently **f** to the treatment.

(iii) The viruses may be unable to enter the cells **g** and so the gene would not be delivered. They might not release the gene **g** once in the cell.

ⓔ **10/15 marks awarded a** The student makes clear s/he is talking about all cancers and then **b** correctly relates percentages to possible numbers, scoring 2 marks. **c** S/he correctly identifies sunburn as a risk factor and then explains it adequately, gaining 2 marks. **d** The student tries to identify a second factor, but it is too imprecise to be credited. **e** These two points gain 2 marks. **f** A correct statement is given and then explained for 2 marks. **g** Two valid points are credited, although the wording referring to 'release' of the gene is a weak expression of gene activation.

ⓔ **Student A scores just 4 marks out of 15, which is a U/E-grade performance. This is largely a result of not supplying the detail and giving answers that are just too vague. You must try to use scientific language wherever possible, and at least make your answers clear. If in doubt, an examiner will mark a poorly expressed answer wrong. You must communicate clearly with the examiner. Student B does this, but also fails to supply detail on some occasions, scoring 10 marks out of 15 — a B-grade answer.**

Knowledge check answers

1 You use a bar chart when the data are categoric; the bars should not touch. You use a histogram when the data are continuous; the bars should touch.

2 The standard deviation takes account of all the data. The range takes account of only the two extreme values and these may be freakish.

3 The error bars overlap in the data for females, so the differences are due to chance. The error bars do not overlap in the data for males, so the differences are probably not due to chance.

4 Because cytosine forms base pairs with guanine, so for every cytosine in the molecule there is a guanine.

5 A gene encodes a polypeptide chain, so genes encoding long polypeptides will have more bases than genes encoding short polypeptides. Also, genes have different numbers and lengths of non-coding base sequences.

6 A DNA molecule, with its associated histones, is too big to pass through the pores in the nuclear envelope. A molecule of mRNA is much smaller and is not bound to histones.

7 A trinucleotide is a molecule formed by the condensation of three nucleotides.

8 (a) 150 (3 bases code for one amino acid). (b) Some of the base sequence might be non-coding — might be an intron or a multiple repeat sequence.

9 23 homologous pairs

10 Half of the nucleotides would be labelled after semi-conservative replication.

11 Only at the end of cytokinesis are there two separate cells. Until that time, the DNA has been held in a single cell that is in the process of dividing.

12 Prokaryotic cells lack a nucleus. Mitosis involves division of a nucleus.

13 Because vincristine prevents the formation of spindles, it will prevent mitosis in cancer cells.

14 There are four chromatids in a homologous pair of chromosomes during meiosis. After one cross-over, two of these will have a new combination of alleles.

15 The effect is the same in both — a reduction in genetic diversity. The founder effect results from a small sample of the original population colonising a new environment. A genetic bottleneck results from a massive reduction in the original population.

16 It encourages outbreeding and so maintains genetic diversity.

17 Prokaryotic DNA is shorter (with fewer loci/genes); it is not linear; it is not bound to histones.

18 The resistance gene arose by a chance mutation; its appearance is not caused by the presence of antibiotic.

19 By removing air/water, with a high carbon dioxide concentration and replacing it with air/water with a high oxygen concentration, it increases the diffusion gradients of these two gases.

20 The gills are supported by water. In air, the gills stick together so that their surface area becomes much too small to enable the fish to survive.

21 The insect's gas exchange system relies on diffusion from the spiracles to the cells. As this distance becomes larger, the efficiency of diffusion becomes less.

22 A leaf respires all the time but photosynthesises during daylight hours. During the daytime, the volume of oxygen released is the difference between that produced during photosynthesis and that used in respiration. Similarly, the volume of carbon dioxide taken up is the difference between that produced by respiration and that used in photosynthesis.

23 The elastic tissue allows the arteries to stretch when blood is forced into them by contraction of the ventricles. It then recoils when the ventricles relax, pushing blood along the arteries.

24 A high water potential has a less negative value than a low water potential. In an examination you can use either term — high water potential or less negative water potential — whichever you find easier.

25 The cell wall is not solid; it has spaces within and between its micelles through which water can move.

26 Root pressure results from the active secretion of ions into the xylem. Lack of oxygen would reduce the rate of aerobic respiration, which produces the ATP used in active transport.

27 Water might be used within the plant, e.g. for hydrolysis reactions.

28 The wavelength of light limits the resolving power of optical microscopes.

29 The symplast pathway is through the cell surface membranes and cytoplasm of cells.

30 An artery consists of different tissues — connective tissue, elastic tissue, muscle tissue and epithelial tissue. A capillary is made of only one type of cell — it is just epithelial tissue.

31 Primary — sequence of amino acids; secondary — folding of chain of amino acids into helix or pleated sheets; tertiary — folding of chain of amino acids into globular structure; quaternary — two or more polypeptide chains in a single protein molecules.

32 From right to left. The blood flows from a high pO_2 in the lungs to a low pO_2 in the respiring tissues (via the left side of the heart).

33 The increased rate of respiration.

34 One that involves two molecules combining with the elimination of a molecule of water.

35 Enzymes usually hydrolyse polymers from their ends. A highly branched molecule has more ends available for enzyme–substrate formation.

36 Mammals and birds

37 They belong to the genus *Homo*.

38 The more closely related the two species, the more similar their DNA base sequences. As a result, there will be more hydrogen bonds formed between complementary base pairs and the higher the temperature needed to break them all.

39 Canopy refers to the layer containing the leaves of the trees; occasional emergents grow above the canopy.

40 Proteins and nucleic acids contain nitrogen.

41 Different types of plant absorb ions at different depths in the soil. Legumes fix nitrogen and, once ploughed back into the soil, add nitrates to the soil.

Index

A

adenine 13–14
agriculture and biodiversity 59
alleles 8, 16
amylopectin 49–50, 51
amylose 49–50, 51
antibiotic-resistance 28–30
antibodies 54, 55
anti-parallel strands 13
apoplast pathway 39
arteries 35, 36
arterioles 35, 36
artificial classifications 51

B

bacteria
   antibiotic resistance 28–30
   effects of antibiotics 28
   prokaryotic cells 14, 53
   selection in 80–81
bactericidal antibiotics 28
bacteriostatic antibiotics 28
bar charts 9–10
base-pairing rule, DNA 13
benign tumours 18
bias 10
binary fission 28–29
binomial 52
biodiversity 56–59, 63–64
blood vessels 35–36
Bohr effect 48–49
bubble potometers 41

C

cancers 18, 82–84
capillaries 35, 36
Casparian strip 39
categoric variation 8
cell cycle 19–20
cell types 43–45

cellular differentiation 45
cellulose 44, 49
cell wall, plants 43–44
centromeres 21, 22
chloroplasts 45
chromatids 18, 21, 22, 23
chromosomes 18, 19, 21–24
circulatory system 35–38
classification 51–53, 74–75
coding strand, DNA 12
complementary bases 13
conjugation 28–29
continuous variation 8
crop rotation 59
crossing over, meiosis 22
cytokinesis 20, 21, 22, 23
cytosine 13–14

D

Darwin, Charles 27
daughter cells 17
deforestation 58
deoxyribose 13, 14
diffusion 30–34
diploid number 18
discontinuous variation 8
DNA 12–17
DNA helicase 19
DNA hybridisation 54–55
DNA polymerase 19
DNA replication 19

E

ecosystem, diversity in 56–57
error bars on graphs 11
eukaryotic cells 14–15, 43, 53
exons 16

F

family 52
farming and biodiversity 59

fish, gas exchange 32–33, 34
founder effect 25

G

gametes 18
gas exchange 32–34
generic names 52
genes 8
   and DNA 12–13
   mutations 16–17
genetic bottlenecks 24, 25–26, 76–77
genetic diversity 24–27
genetic material, passing on 17–24, 78–79
genetic variation 8–12
genus 52
gills of fish 32
glycogen 11, 50, 51
guanine 13–14
guard cells 44–45

H

haemoglobin 47–49, 68–70
haploid number 18
histograms 9–10
homologous chromosomes 18, 22–23
horizontal gene transmission 28, 29

I

in-breeding 26
index of diversity 56–57
insects, gas exchange 33, 34
interspecific variation 8
intraspecific variation 8
introns 13, 16

K

kingdoms 52–53